»On the 6" section of wire, sand off all of the enamel except for the top 2".

5. BEND THE WIRE

»Bend the 1½" non-sanded end of the long wire into a small loop. On the other end, bend the wire into a loop a little smaller than your cat toy. This is the pendulum.

»Bend the sanded end of the short wire into a quarter-sized loop. Next, bend about half of the wire into a large circle, parallel to the first circle. This will be the base.

»Create a small hook on the non-sanded end of the remaining wire and then bend the wire up from the base so the hook sits directly above the small loop.

6. ATTACH WIRES AND ADJUST

»Run the small loop in the long wire up through the quarter-sized hole in the stand, and attach it to the hook you bent into the stand. Make sure that it dangles in the center of the hole in the base so the sanded portion can make contact with the ring it's passing through.

»You may need to bend the base section a little to make sure everything is aligned. The insulation on the hook and inside the small loop on the pendulum should keep them from being electrically connected, but when the pendulum swings and touches the side of the ring, a connection will be made.

7. ATTACH THE SIGNAL WIRES

»On each of the 2' red and black hookup wires, strip about 2" of insulation from one end and ¼" from the other.

»Wrap the 2" of stripped black wire around a sanded portion of the base. Make sure it is tight, and then wrap the wire in tape to insulate it and keep it in place.

»Wrap and tape the 2" section of stripped red wire onto the long pendulum wire. Make sure the tape doesn't insulate the pendulum from the ring.

8. WIRE YOUR ARDUINO

»Plug the free end of your black wire into ground on your Edison Arduino Breakout and the red wire into Pin 10. Jump from any 5v pin over to Pin 10 with the 10k resistor — this will act as a pull up resistor so we can detect the switch closing.

9. SET UP ARDUINO TWEET

»Tweeting from Arduino platforms couldn't be easier with the Tweet Library for Arduino (arduino-tweet.appspot.com). This web app will do the heavy lifting for you by authenticating to Twitter and sending the message.

»To access the web app, all you need is your API key. Follow the first two steps in their setup process to get everything you need to send tweets. I created a special Twitter account for my cat toy so those following me wouldn't be distracted by the tweets coming from it.

10. GET THE CODE

»Download the IOCat.ino sketch at github.com/MattStultz/IOCat.

»Add your Twitter Library for Arduino key and modify the tweet message to include the link to your Edison streaming server.

»Then, upload the code to your Edison module via the Arduino IDE.

11. PUTTING IT ALL TOGETHER

»Plug the USB webcam (I recommend a Logitech C270, but any UVC-compatible webcam works fine) and power cord into your Edison and connect to it via SSH. Navigate to the directory "Edi-Cam/Web/Server" and run the command "**node server.js &**". The "**&**" will allow the code to continue running even after you disconnect your SSH session.

»Tape or clamp the pendulum base to the edge of a table.

»Attach one of your cat's favorite toys to the end of the pendulum with tape or a light thread (if kitty gets too aggressive, you don't want the whole thing coming down).

»Sneak away and wait for the tweets to start rolling in.

TAKING IT FURTHER

The link from this device will only be visible to you if you are on the same network as the device. If you would like to open it up to the rest of the world, you can do so by creating a dynamic DNS address and using a little port forwarding magic. Now, sit back and feed the internet more cat content as your kitty bats away at its toy. ⊘

Hep Svadja & Matt Stultz

CONTENTS

COLUMNS

Nick Normal

10

FEATURES

18

ON THE COVER:
Adam Savage's Navy Mark IV Mercury Space Helmet.
Photo: Hep Svadja

26

SPECIAL SECTION
CITIZEN SPACE

Issue No. 47, October/November 2015. *Make:* (ISSN 1556-2336) is published bimonthly by Maker Media, Inc. in the months of January, March, May, July, September, and November. Maker Media is located at 1160 Battery Street, Suite 125, San Francisco, CA 94111, 877-306-6253. SUBSCRIPTIONS: Send all subscription requests to *Make:*, P.O. Box 17046, North Hollywood, CA 91615-9588 or subscribe online at makezine.com/offer or via phone at (866) 289-8847 (U.S. and Canada); all other countries call (818) 487-2037. Subscriptions are available for $34.95 for 1 year (6 issues) in the United States; in Canada: $39.95 USD; all other countries: $49.95 USD. Periodicals Postage Paid at Sebastopol, CA, and at additional mailing offices. POSTMASTER: Send address changes to *Make:*, P.O. Box 17046, North Hollywood, CA 91615-9588. Canada Post Publications Mail Agreement Number 41129568. CANADA POSTMASTER: Send address changes to: Maker Media, PO Box 456, Niagara Falls, ON L2E 6V2

28

58

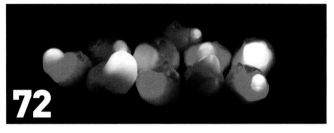

72

76

80

62

Nate Van Dyke

Matthew Billington

EXECUTIVE CHAIRMAN
Dale Dougherty
dale@makermedia.com

CEO
Gregg Brockway
gregg@makermedia.com

CFO
Todd Sotkiewicz
todd@makermedia.com

"Exploration is in our nature. We began as wanderers, and we are wanderers still. We have lingered long enough on the shores of the cosmic ocean. We are ready at last to set sail for the stars." — Carl Sagan, *Cosmos*

EDITOR IN CHIEF
Rafe Needleman
rafe@makezine.com

EDITORIAL

EXECUTIVE EDITOR
Mike Senese
mike@makermedia.com

PRODUCTION MANAGER
Elise Tarkman

COMMUNITY EDITOR
Caleb Kraft
caleb@makermedia.com

PROJECTS EDITORS
Keith Hammond
khammond@makermedia.com
Donald Bell
donald@makermedia.com

TECHNICAL EDITORS
David Scheltema
Jordan Bunker

EDITOR
Nathan Hurst

EDITORIAL ASSISTANT
Craig Couden

COPY EDITOR
Laurie Barton

LAB MANAGER
Marty Marfin

LAB PROJECT EDITOR
Emily Coker

EDITORIAL INTERNS
Sophia Smith
Nicole Smith

CREATIVE DIRECTOR
Jason Babler
jbabler@makezine.com

DESIGN,
PHOTOGRAPHY
& VIDEO

ART DIRECTOR
Juliann Brown

DESIGNER
Jim Burke

PHOTOGRAPHER
Hep Svadja

VIDEO PRODUCER
Tyler Winegarner

VIDEOGRAPHER
Nat Wilson-Heckathorn

MAKEZINE.COM

DESIGN TEAM
Beate Fritsch
Eric Argel
Josh Wright

WEB DEVELOPMENT TEAM
Clair Whitmer
Bill Olson
David Beauchamp
Rich Haynie
Matt Abernathy

VICE PRESIDENT
Sherry Huss
sherry@makermedia.com

SALES
& ADVERTISING

SENIOR SALES
MANAGER
Katie D. Kunde
katie@makermedia.com

SALES MANAGERS
Cecily Benzon
cbenzon@makermedia.com
Brigitte Kunde
brigitte@makermedia.com

STRATEGIC
PARTNERSHIPS
Angela Ames

CLIENT SERVICES
MANAGERS
Margaux Ryndak

COMMERCE

GENERAL MANAGER
OF COMMERCE
Sonia Wong

RETAIL CHANNEL
DIRECTOR
Kirk Matsuo

ASSOCIATE PRODUCER
Arianna Black

E-COMMERCE MANAGER
Michele Van Ruiten

MARKETING

VICE PRESIDENT
OF CORPORATE
MARKETING
Vickie Welch
vwelch@makermedia.com

DIRECTOR OF
CUSTOMER
ACQUISITION
Patrick McCarthy
patrick@makermedia.com

CUSTOMER RETENTION
MANAGER
**Heather Harmon
Cochran**
heatherh@makermedia.com

MARKETING PROGRAMS
MANAGER
Suzanne Huston
suzanne@makermedia.com

DIGITAL MARKETING
COMMUNICATIONS
MANAGER
Brita Muller
brita@makermedia.com

MARKETING EVENTS
MANAGER
Courtney Lentz
courtney@makermedia.com

SOCIAL MEDIA
MARKETING MANAGER
Jessie Wu
jessie@makermedia.com

BOOKS

PUBLISHER
Brian Jepson

EDITORS
Patrick Di Justo
**Anna Kaziunas
France**

MAKER FAIRE

PRODUCER
Louise Glasgow

PROGRAM DIRECTOR
Sabrina Merlo

MARKETING & PR
**Bridgette
Vanderlaan**

SPONSOR RELATIONS
MANAGER
Miranda Mota
miranda@makermedia.com

CUSTOM
PROGRAMS

DIRECTOR
Michelle Hlubinka

CUSTOMER
SERVICE

CUSTOMER SERVICE
REPRESENTATIVES
Ryan Austin
Camille Martinez

Manage your account online, including change of address:
makezine.com/account
866-289-8847 toll-free in U.S. and Canada
818-487-2037,
5 a.m.–5 p.m., PST
cs@readerservices makezine.com

PUBLISHED BY

MAKER MEDIA, INC.
Dale Dougherty

Copyright © 2015
Maker Media, Inc.
All rights reserved.
Reproduction without permission is prohibited.
Printed in the USA by Schumann Printers, Inc.

CONTRIBUTING EDITORS
Stuart Deutsch, William Gurstelle, Nick Normal, Charles Platt, Matt Stultz

CONTRIBUTING WRITERS
Cabe Atwell, John Baichtal, Signe Brewster, Emma Chapman, Eric Chu, Jeremy Cook, Michael Curry, DC Denison, Kranti Gunthoti, Samatha Kranthijanya, Ben Krasnow, Rik Kretzinger, David Lang, Brandon Lawler, Forrest M. Mims III, Dan Rasmussen, Matthew F. Reyes, Warren Simons, Andrew Terranova, Tested.com, Mike Westerfield

Comments may be sent to:
editor@makezine.com

Visit us online:
makezine.com

Follow us on Twitter:
@make @makerfaire @craft @makershed

On Google+:
google.com/+make

On Facebook:
makemagazine

CONTRIBUTING ARTISTS
Matthew Billington, Nick Dragotta, Bob Knetzger, Rob Nance, Andrew J. Nilsen, Charles Platt, James Provost, John Thomas, Nate Van Dyke

CONTRIBUTING DESIGNER
Boni Uzilevsky

ONLINE CONTRIBUTORS
Gareth Branwyn, Jon Christian, Marc de Vinck, Jimmy DiResta, Miriam Engle, Adam Flaherty, Matt Freund, Roger Garrett, Karen Hickman, Grady Hillhouse, Meredith Lee, George LeVines, Goli Mohammadi, Agnes Niewiadomski, Jenn Nowicki, Quentin Oliver, Haley Pierson-Cox, Robert Ponzio, Mancil Russell, Andrew Salamone, Eric Schemplefeng, Ward Silver, Theron Sturgess, Igor Vichikov, Madison Worthy

MAKER MEDIA LAB INTERNS
Samuel DeRose, Adam Lukasik, Cameron Mira, Willy Nicholas, Jose Santos, Eileen Welch, Chelsea Wirth

CONTRIBUTORS

What area of the world — or beyond — would you most like to explore, and how?

Signe Brewster
San Francisco, California [Hope from Above]
I'm eager to see if there is any life on Europa, so I would explore its ocean in a submarine.

Matthew F. Reyes
Redwood City, California [The NASA Challenges]
I love exploring harsh geological features in the hunt for strange forms of life. On Earth, Mars, or the moon, the best way would obviously be by motorbike.

James Provost
Toronto, Ontario [Illustrator, Near-Space Photography]
Maps of Mars are about 250 times more detailed than maps of the Earth's ocean floors. I'd use a swarm of autonomous bots to map the sea floor in higher resolution.

Kranti Gunthoti & Samatha Kranthijanya
Batavia, Illinois [Cosmic Ray Detector]
We would like to explore moons of Jupiter and nearby planets with a home-made telescope we are planning to build.

Michael Curry
Kansas City, Missouri [Ye Olde Brushless Motor]
People occupy a small percentage of Earth's surface, and it is shrinking. I'm fascinated with robotically exploring the places we leave behind.

PRINTED WITH SOY INK

Congratulations!

To Our
Challenge Finalists!

View the final entries and cast your vote
for People's Choice Award at Maker Faire NYC!

As the first round of the challenge comes to a close,
we encourage you to stay tuned for the upcoming rounds
that promise to be just as exciting!

americamakes.us/challenge

PRIZES BY

CENTENNIAL
CHALLENGES

America Makes

Make:

Excitement for InMoov and its Creative Community

Hep Svadja

>> Your article about the new progress in making a printable humanoid robot is amazing. It is astonishing how much the community has advanced in so many fields of science to be able to create something like InMoov. It is always portrayed in movies and books as tech far into the future, but in truth it might be available very soon. I think that this is a very exciting prospect, and I also believe that this technology can be used in many more ways than just to make a robot. While creating an artificial body that can perform all the same actions that a human can is astounding, we can eventually use this science for other things — things that we are already working on, such as artificial limbs that act the same as the real ones, or to simulate how an actual human body would do something.

–Nicholas Holly, Woodstock, Georgia

>> ["InMoov Around the World" from *Make:* Volume 45] is a great feature. I really like the InMoov design and good to see it's becoming ever more popular. The ideas that are implemented in the InMoovs featured just goes to show what can be done with this platform with some clever lateral thinking and hard work. This is a great showcase of the excellent work done by all who are featured in this article. Well done everyone.

–Steve Gibbs, Mitcham, Surrey, United Kingdom

BRINGING "TOYS" INTO THE CLASSROOM

Thank you for the amazing variety of articles in each issue of *Make:*. I teach sixth grade general science, and although I can appreciate many of your articles, I particularly am drawn to "Toy Inventor's Notebook" by Bob Knetzger. I teach well over 100 students, and therefore many of the more "high tech" projects are more difficult for me to use with students due to the materials/complexity/cost. Bob's articles tend to be appropriate for large numbers of students. My sixth graders and I want to thank you for including the "Toy Inventor's Notebook" in *Make:* We are big fans!

–Doug Stith, Londonderry Middle School, Londonderry, New Hampshire

A TIP IN RESPONSE TO SOLDERING IRON TIPS FROM *MAKE:* VOLUME 45

Here's another tip (sorry): A small conical tip is nice for getting into small places, like soldering the pins on a QFP package, for example. Change to a large chisel, though, before trying to solder anything connected to a large ground plane. The tiny tips can't deliver enough heat fast enough, before the large ground area sucks it away.

–Antron Argaiv, via the web

IN RESPONSE TO "TO BUILD A BETTER ROBOT" FROM *MAKE:* VOLUME 45

I mentor a middle school robotics team and find your article interesting. I sincerely hope that none of the girls on my team have experienced any sexism and have never seen any indication that they have. We work hard to treat all of the students fairly and to elevate the girls and boys in equal numbers to leadership positions. In my personal experience, the girls are at least as capable as the boys and tend to be more enthusiastic and make excellent leaders.

–Steve K., via the web

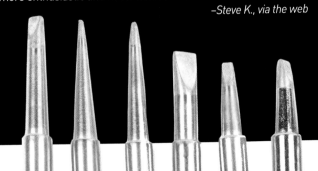

Vik Orenstein

A Construction Set for the 21st Century

BY DALE DOUGHERTY, founder and Executive Chairman of Maker Media

THE MAKER MOVEMENT IS HAVING A BREAKOUT YEAR IN CHINA. I could see it at Maker Faire Shenzhen, a large, bustling celebration this July that was driven in part by government sponsorship. SEEED Studio's Kevin Lau, the executive producer of Maker Faire Shenzhen who organized hundreds of Makers across four city blocks, said it was 10 times larger than 2014.

China's premier, Li Kequiang, has focused on economic reform. During a visit to a Shenzhen Makerspace in January, he declared that "Makers with creative ideas should be helped to set up their own businesses." The deputy mayor of Shenzhen also spoke about this emerging opportunity: "Shenzhen is to be a city built for Makers." The government wants to put together an entrepreneurial ecosystem for Chinese Makers that supports startups and small businesses, and the prototype for a new generation of entrepreneurs could be Jasen Wang of Makeblock.

Wang is a soft-spoken engineer with a gleam in his eye. Five years ago, after finishing his master's degree, he came to Shenzhen to start a hardware company. In 2011 he founded Makeblock, a construction kit that looks like a wholly reimagined Erector set, and was accepted into the first class of companies in Shenzhen-based incubator HAX Accelerator. He ran his first Kickstarter campaign in 2013, raising $185,000. Last year, Wang raised $6 million from Sequoia Capital and this year expanded his team from 10 to 90 people, adding expertise in manufacturing, software development, and design.

Makeblock consists of aluminum beams, mechanical components, and electronic modules, and features an Arduino-compatible controller and standard sensors. "We combine a lot of new technology, including open-source technology," explains Wang. Makeblock offers kits for building a musical robot, an XY-Plotter, and even your own 3D printer. The products are well designed, colorfully packaged, and competitively priced.

Wang showed me a new Makeblock product that he was excited about: Mbot, an educational robot for kids, offers a Scratch-like programming environment to control an Arduino-compatible robot. The Kickstarter campaign for Mbot wrapped up in May, raising $285,000

from 2,500 backers. Wang believes the $75 kit is "affordable for every kid so they can learn both robotics and programming." Mbot comes in two colors, blue and pink, which will please some and infuriate others.

Maker Faire Shenzhen was full of Makers who had run successful Kickstarters, not only to generate capital but also to market their products. I would bet that most of their backers are not from China. What impressed me the most, however, was how quickly they can expect to deliver a product after their campaigns are funded, by tapping into Shenzhen's manufacturing capabilities. Wang planned to ship Mbot a month or two after the campaign closed. Most Americans who run hardware campaigns on Kickstarter take around 18 months to manufacture and fulfill a product — if they are lucky.

Wang says that Maker Faire Shenzhen exposed more Chinese to the Maker Movement, and adds that they sold a lot of Mbots during the Faire. "We have a lot of new customers because parents who never heard of 'Maker' now know what 'Maker' means, and they understand that this is important for their kids and themselves."

As part of Shenzhen's Maker Week, Makeblock held a 48-hour robotics competition. The theme was "Lyre-Playing, Chess, Calligraphy, and Painting," and a dozen teams were brought to Shenzhen and provided with all the Makeblock components they needed to build a project. Teams came from as far as Utah and Italy, and several built variations of drawbots — one was a watercolor bot, another splattered paint in creative ways. The winner was an Italian team called /dev/null, which built a Light Saber Chessbot, a two-foot tall chess piece whose movement can be directed by a laser pointer. The teams demonstrated ingenuity and teamwork, and that you can build almost anything with Makeblock.

The original Erector set was created by W.S. Gilbert who believed that if children are given the right tools, they will educate themselves. Gilbert was inspired by the girders hoisted in the air to build skyscrapers in New York City in the early 20th century. Wang's Makeblock, which integrates physical and digital modules, is no doubt inspired by Makers who are furiously building things in Shenzhen in the 21st century. ✐

Makeblock

MADE ON EARTH

The world of backyard technology

Know a project that would be perfect for Made on Earth?
Email us: editor@makezine.com

GALACTIC GRACE

CELESTIAL-MECHANICA.ORG

As a child, **Jessika Welz** was enthralled by a scene in the popular Jim Henson movie *The Dark Crystal*, where the entire set is filled with a massive mechanical solar system called an orrery.

Typically, these functional models are small, precise clockwork designs, but what stood out to Welz was the massive grandeur of the movie set piece. The imagery of the mechanical precision mixed with the gargantuan proportions stuck with her for years, ultimately resurfacing when she decided to build an art piece called *Celestial Mechanica* for Burning Man.

With the help of a devoted team of people over several months, Welz's vision was brought to life. Each piece was meticulously crafted from bare metal specifically for the purpose of inspiring that same awe she felt as a child.

The orrery's 40-foot-wide span of orbit, and the 6-foot-diameter, propane-powered, flaming sun at the core easily brings that same grandeur to Welz's design. You can feel the heat on your face as you stand outside the farthest orbit and observe the custom-made gears rotating each planet and moon within a mathematically correct model.

"I was going to make an entire fantasy solar system, but I started looking into our own solar system and it's really quite fascinating on its own," says Welz.

— *Caleb Kraft*

ROCKIN' ROBOTS KOLJAKUGLER.COM

Kolja Kugler loves to make big sculptures out of scrap metal, which satisfied him until he discovered pneumatic cylinders in 1999. With air power, his sculptures transformed from static pieces held together with blood, sweat, and welding wire, into something much more. "I added the cylinders to the face I was just building, and then, suddenly, the face came alive," he says.

This face eventually morphed into a full body, now known as "Sir Elton Junk." This robotic sculpture is the manager of Kugler's Germany-based *One Love Machine Band*, which consists of a drummer, a bass guitarist, and a bird-shaped flutist. The band, although more than capable of excellent robo-music, focuses more on sculpture and expression. Kugler gets inspiration from many sources, including *The*

Muppet Show, which seems apparent in the way his robots look and carry themselves.

The robotic band's show continues to evolve. Kugler's considered adding human artists, or even building other robot "acts," like making a pneumatic tiger jump through a hoop. We hope he'll consider setting that hoop on fire.

— *Jeremy Cook*

Maxwell Stephen Duryea

David Lavoie

MARITIME MARVELS BUILDEROFSHIPS.COM

Building a ship in a bottle is typically accomplished via one of two methods — building the ship outside with hinged masts that are raised using strings once inside the bottle, or the more difficult and time-consuming method of using long-handled tools to build it directly inside.

David Lavoie, who is self-taught, uses a combination of both methods for his illuminated vessels. "I carve and build the ship outside the bottle, then I disassemble it and rebuild it piece by piece inside the bottle," he says. "The fiber optics, which are bundled at the end of the mast and go through the deck do not allow enough clearance to fully collapse the masts. I've been building my ships piece by piece within the bottle."

The fiber optic light also causes diffuse reflection and scattering, which can transmit in random directions, especially when the fiber is bent or angled. The finished build takes on an eerie, glowing brilliance thanks to the lights and simulated rough seas.
— *Cabe Atwell*

OPEN

WRITTEN BY DAVID LANG

EXPLORERS
OF THE UNIVERSE

GEECs.com

MAKERS ARE REINVENTING AND REIMAGINING THE PROCESS OF DISCOVERY

DAVID LANG
is the co-founder
of OpenROV and
OpenExplorer. He
is also a TED Senior
Fellow, a member
of the NOAA Ocean
Exploration Advisory
Board, and the author
of *Zero to Maker*.

Erika Bergman,
*Girls Underwater
Robot Camps*

THE HISTORY OF SCIENCE AND EXPLORATION IS A MAKER STORY. Our tools, and the people who shape them, have always determined the questions we ask, the horizons we pursue, and the answers we discover. From microscopes to telescopes, submarines to spaceships, it's always taken a Maker (or team of Makers) to push the limits of what we know.

Over the past century, though, exploration became inaccessible. A distinction between "professional" and "amateur" science created a barrier where one had never existed. Historically, curiosity was pursued by those who felt inspired, and they built whatever they needed to get the job done. However, well-meaning and effective systems — federal grant funding, conferences, and journals — have made the pursuit of knowledge an exclusive activity for those in the ivory towers.

Over the past few years, thanks to the Maker Movement, the tools for science, exploration, and conservation have become more powerful and more obtainable. Cheap components, open standards, and connected enthusiasm are driving Makers to reinvent and reimagine the process of discovery.

The professional science community is beginning to sit up and take notice. They don't see it as a threat, but an opportunity. Maker-style science is not replacing any traditional efforts, it's just more: experiments, expeditions, and possibilities. It's enabling a whole new genre of question-askers. It's both surprising and thrilling.

We've connected with a new generation of Maker-explorers to hear their stories and get inspired by their projects. Around the world, amazing science is happening. Makers are reclaiming their place as explorers of the universe.

Hardshell Labs

GEECs.com

LASER TAG FOR TORTOISES

Tim Shields, *Hardshell Labs*

hardshelllabs.com

A field biologist for the Bureau of Land Management, Tim Shields had developed a frustration with the ravens that were decimating the desert tortoises he spent his life studying in the Mojave Desert.

Given the chance, a raven will peck a hole in the shell of a young tortoise and eat it. As people moved into and around the Mojave, they have modified the environment, bringing more water, shade, roosts, and food. The ravens' numbers grew alongside — a 1,000% increase from 1975 to 1995 — and the tortoise population plummeted. Shields responded with high-tech measures and started chasing off ravens with powerful laser pointers.

But even the most dedicated field biologist can't stand in the desert shining lasers at ravens for weeks. So Shields' endeavor evolved into building laser-equipped buggies to protect his favorite species. So far, it's working.

"We're currently using ground-based rovers to observe desert tortoises without disturbing them, and equipping them with raven repulsion devices to allow operators to drive these predators away from vulnerable juvenile tortoises," says Shields. He is even working on ways to gamify the project, which is licensed partially under Creative Commons, so remote operators can control the lasers. "Our goal is to engage active conservationists who currently lack the capacity to witness the wonders of the natural world or to do work to intervene ecologically," he says.

DIVING DEEP

Erika Bergman, *Girls Underwater Robot Camps*

openexplorer.com/expedition/girlsunderwaterrobotcamp

Submarine pilot and National Geographic Young Explorer Erika Bergman is helping inspire the next generation of researchers and scientists. She is the co-founder of the Girls Engineering and Exploration Counselors, which organizes the Girls Underwater Robot Camps to provide teenage girls engineering and robotics experience.

Since 2014, Bergman's team has run workshops, speaking engagements, and pilot camps. The camps themselves focus around the construction of an OpenROV submersible, followed by planning and running a micro-expedition in the field. The completed craft are then made available for school projects and expeditions.

Bergman put herself through college as a marine diesel mechanic and steam boat fireman, and taught herself to pilot a submarine on the job in 2010. She now drives various research vessels, including a five-person sub that is capable of diving to 300 meters. It's those experiences that have motivated her outreach endeavors.

"I envisioned the impact of bringing young women along on expeditions, but I couldn't physically take girls with me on oceanographic research ships," she explains. "I could however bring something they were deeply connected to, something they had built with their own hands that would connect future ocean explorers. Building an underwater robot together was an opportunity to deliver their spirit and hard work into the hands of other girls across continents."

makezine.com/open-explorers

Hannah Perner-Wilson

MADAGASCAR MAKERSPACE

Andrew Quitmeyer and Hannah Perner-Wilson, *Hiking Hacks*
openexplorer.com/expedition/ disseminationlabmadagascar

While on a mission to find a rumored ant species in Madagascar, Andrew Quitmeyer and Hannah Perner-Wilson needed a place to put together a sensor triggered by ants, so they built a Makerspace in a tent.

The trip was one of a series of Hiking Hacks, which take Quitmeyer and Perner-Wilson around the world, building and sharing tools and sensors that are specialized to the ecosystem where they're deployed. In Madagascar, they used new technology to document and disseminate biological field research and projects (the ant sensor and other projects are up on Instructables). That's tricky in remote locations, so they worked to build a field-friendly, mobile "hacking center" that acts as a communal design space and also protects equipment from the elements.

"I realized that when building technology designed for interacting with nature, it might make sense to build technology *in* nature," says Quitmeyer, a digital media Ph.D. student at Georgia Tech.

If they can scale the Makerspace idea, it could help get more people building their tools in the places they'll be used.

"The jungle is an amazing setting for building electronics," says Perner-Wilson, who has a background in e-textiles and DIY electronics. "All the plants and animals growing and living around us were extremely useful building materials and inspiration for project ideas."

DETECTING ILLEGAL CHAINSAWS

Topher White, *Rainforest Connection* **rfcx.org**

On an ecotourist jungle trek, Topher White had a startling encounter. "One day, our group was taken on a guided walking tour in the reserve, and within five minutes walk from the ranger's headquarters, we came upon a group of illegal loggers," he explains. "They fled upon our arrival, but the problem was evident: Even within relatively short distances, rangers had no real-time awareness of destructive activity in their reserve."

White, a physicist and engineer, used that incident as inspiration to develop a device that repurposes old cellphones into rainforest monitoring systems. He launched a Kickstarter to build the systems, which resulted in one of the most successful conservation projects on the site. Now his resulting organization, Rainforest Connection, is placing the apparatus in forests worldwide.

Rainforest Connection

One of the keys to the system is the encroachment of technology that it is fighting to protect the forest from. "I noticed that even in the jungle, bereft of electricity and perhaps a hundred miles from the nearest road, there was quite reliable cell service. And the populations there had come to rely upon it," White says. Sensitive microphones wired to the phones listen for the sound waves of chainsaws or vehicle traffic, using the cellphone network to send a text message alert when detected. Solar panels attenuated to the speckles of sunlight that peek through the forest canopy keep batteries charged, and the arrays are placed high on trees to keep them above detection.

"It's always an adventure," White says. "Climbing 200-foot trees on a daily basis and backpacking through the jungle with Amazon warriors is hardly the type of work that I'd expected to be doing with a physics degree, but these days it makes total sense to me."

Hope
FROM ABOVE

Written by Signe Brewster

Designing drones to deliver aid to Syria

Hep Svadja and Getty Images

SIGNE BREWSTER
is a San Francisco-based science and technology journalist who covers robotics, drones, 3D printing, and the Maker Movement.

ON MARCH 16, BARRELS OF CHLORINE GAS RAINED DOWN ON THE TOWN OF SARMIN IN NORTHERN SYRIA, KILLING SIX AND WOUNDING MANY MORE — just one of many horrific chemical attacks in the civil war that has consumed the country.

"Sarmin isn't far from the border but the border is closed to all traffic," Sasha Ghosh-Siminoff, president of Syrian aid organization People Demand Change, texted his Stanford University-based friend Mark Jacobsen, four hours after the attack. "If your planes were ready, you could have flown in emergency medicine and gear."

The planes Ghosh-Siminoff was referring to are drones, built expressly for this purpose. Jacobsen is the executive director of Uplift Aeronautics, a nonprofit which hopes to deliver essential medical supplies, food, and other cargo to Syrians via its Syria Airlift Project. Syria recently closed its border to foreign aid, and any planes that attempt to fly over the country run a high chance of being shot down. Uplift has a different plan: Fleets of drones that could swoop in by night, undetected by human eyes or radar.

DOING MORE

Jacobsen, who is pursuing a Ph.D. in political science, was in Istanbul about a year ago with a group of academics when a heated discussion broke out about international intervention in the Syrian war. Since 2011's Arab Spring, when activists came together to protest president Bashar al-Assad and his government, at least 200,000 people have died there. More than 10,000 were children. A lack of medical care and food are among the government's weapons against its own people.

Person after person at the gathering asked the same question: Why isn't more being done? Jacobsen, a former Air Force cargo pilot, explained to one attendee that you simply can't fly a cargo plane into such an unpredictable place. It's impossible.

He went back to his hotel that night feeling guilty. It didn't seem like a good enough answer. While speaking with his colleagues, he became fixated on the idea of sending in large numbers of

MECHANISM FOR *Remote Cargo Deployment*

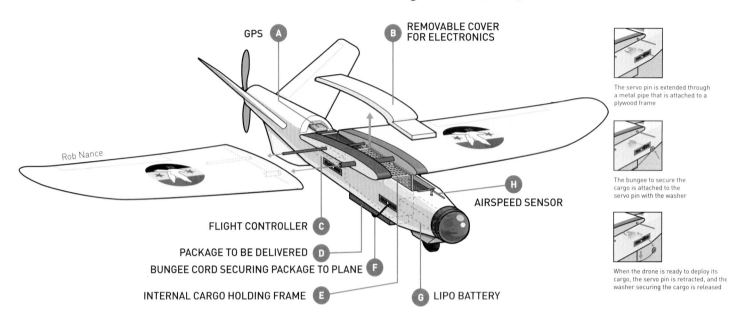

GPS **A**

B REMOVABLE COVER FOR ELECTRONICS

Rob Nance

AIRSPEED SENSOR **H**

FLIGHT CONTROLLER **C**

PACKAGE TO BE DELIVERED **D**

BUNGEE CORD SECURING PACKAGE TO PLANE **F**

INTERNAL CARGO HOLDING FRAME **E**

G LIPO BATTERY

The servo pin is extended through a metal pipe that is attached to a plywood frame

The bungee to secure the cargo is attached to the servo pin with the washer

When the drone is ready to deploy its cargo, the servo pin is retracted, and the washer securing the cargo is released

---

packages — perhaps via drone. He took out a notebook at around 2 or 3 a.m., the hope of sleep long forgotten.

"It seemed like I was onto something with the idea of swarming small packets, but I didn't really know what technology could do that, whether it would be quadcopters or planes or catapults or anything else. Balloons?" Jacobsen says. "I was just trying to lay out everything I could think of."

Uplift Aeronautics and the Syria Airlift Project were eventually born, and today Jacobsen and a group of volunteers are busy flying prototype drones. Their plan is to fly hundreds over the border from a neighboring country on missions chosen by aid partners such as People Demand Change. Each can carry only a few pounds of supplies, but their small size makes them untrackable by radar and dispensable. If a chlorine bomb explodes, medicine-carrying drones can be there in an hour, as opposed to days — or never.

Uplift plans to train Syrian refugees and other people on the ground to fly and repair the drones. Its first destination would be Aleppo, Syria's largest city. The war has hit it hard. Hunger and disease are common.

The drones would take about a half hour to fly to Aleppo. Instead of touching down, they would drop their cargo in a small box attached to a parachute. Then they would return. Back at the launch base, the location of which would likely shift from day to day, volunteers could switch out the battery, load new cargo, and launch again within minutes.

COMPLICATIONS

Flying anything, let alone hundreds of drones, into a country without permission is a breach of international law. Current sanctions bar sending U.S. goods into the country. In extreme times like these, exceptions can be granted, but they depend on various government channels.

Jacobsen isn't exactly sure how Uplift will secure an OK from the U.S., though he has initiated conversations with officials. The drones will likely have to be approved by the U.S. Treasury and international agreements, and will need to comply with arms regulations and counter-terrorism laws.

The local approval process could also be messy. Uplift will need to schedule audiences with governments in countries

bordering Syria, such as Turkey or Jordan. They will need to prove that the drones will be safe and beneficial. The recent election in Turkey, and the country's air strikes within Syria add a new layer of complexity.

"In some ways, negotiating with the armed groups and the people inside Syria is easier than the Turkish governments," Ghosh-Siminoff says. "It's really difficult to navigate that bureaucracy and know you're in the clear and not running afoul of some archaic rule."

Inside Syria, it's actually the groups fighting Assad that would be most likely to shoot down a drone. Currently, it's the resistance that occupies the ground between Uplift's launch site and Aleppo. But if Uplift can demonstrate the planes are for aid, and will not interfere with the opposition's efforts, Ghosh-Siminoff said there should not be a problem convincing the locals to let them pass.

In a country strapped for resources, a scenario could arise where troops start capturing drones to use for their own purposes. Uplift thought of that. The drones are equipped with a self-destruct device designed to fry their navigation system if

CHART FOR *Drone Swarm Flight Plan*

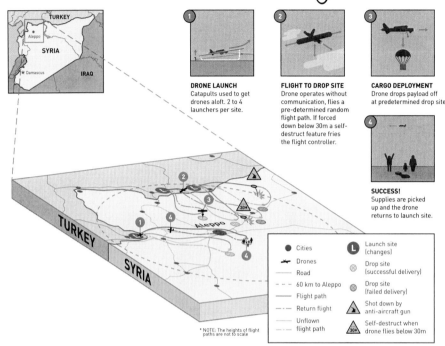

DRONE LAUNCH
Catapults used to get drones aloft. 2 to 4 launchers per site.

FLIGHT TO DROP SITE
Drone operates without communication, flies a pre-determined random flight path. If forced down below 30m a self-destruct feature fries the flight controller.

CARGO DEPLOYMENT
Drone drops payload off at predetermined drop site.

SUCCESS!
Supplies are picked up and the drone returns to launch site.

● Cities	Ⓛ Launch site (changes)
✈ Drones	⊗ Drop site (successful delivery)
— Road	⊗ Drop site (failed delivery)
- - 60 km to Aleppo	⚠ Shot down by anti-aircraft gun
— Flight path	⚠ Self-destruct when drone flies below 30m
-·- Return flight	
···· Unflown flight path	

* NOTE: The heights of flight paths are not to scale

Michael Taylor, with Uplift Aeronautics' Waliid drone mounted on a home-built PVC launcher

When the cargo bay releases a shipment, the weight of the box pulls a parachute, made from plastic bags, out of the drone

Hep Svadja

they fly too close to the ground anywhere but at the launch site. Any drone that gets too low will never be remotely pilotable again.

"We're not planning to talk to them at all once they leave the takeoff area," says Jacobsen about his aircraft. "Routes will be preprogrammed. Our custom firmware on the plane actually plugs its ears and stops listening to incoming messages while in Syrian airspace, which should make it considerably harder to hack."

A TEAM OF VOLUNTEERS

On a hot, cloudless day in April, Jacobsen and four volunteers gathered at Stanford University's Lake Lagunita. Engineer Michael Taylor, a Ph.D. candidate in electrical engineering, led two other volunteers through setting up the drone launcher on the lake bed, which California's drought has turned into a grassy field.

On a porch overlooking the lake, Jacobsen assembled and tested his "Waliid" drone. He ran new volunteer Stuart Ginn, a medical resident still clad in scrubs, through the plane's software and pre-flight protocol.

Made of foam and held together by tape, the drone is not visually impressive. It's shaped like a plane, which allows it to fly for an hour instead of minutes, unlike the quadcopters that have taken over the consumer market. Its wingspan measures 5' 7" and is decorated in black, green, and red — the colors of the Syrian flag.

Back in the field, Taylor and aeronautics and astronautics Ph.D. student Heather Kline had completed the launcher — a 7-foot-long PVC pipe skeleton that guides the drone into the air. Tomoki Eto, a mechanical engineering undergraduate and experienced drone pilot, anchors a bungee line to the ground several hundred feet away, stretches it to the launcher, and attaches it. Upon release, the bungee will fling the UAV into the air.

The team consists of five volunteers and anywhere from 15 to 50 outside participants. The engineering core resides at Stanford, but people all over the world are contributing to its design and deployment. It's been an informal collaboration via email, Skype, and Dropbox, but Uplift plans to release as much open-source material as possible via Github.

Like many of the volunteers, lead engineer Brandon Fetroe got involved with the project after hearing about it through Stanford's UAV club. A mechanical engineering Ph.D. student, Fetroe has been flying R/C planes since he was 12. He described his expertise as a little bit of everything — something that holds true throughout the Uplift team. Ginn, for example, was once a commercial pilot; he's now helping reach out to medical NGOs. And Jacobsen is leveraging his international contacts and friends in the U.S. government from his days in the Air Force.

BREAKING NEW GROUND — AND AIR

Interest in using drones altruistically is high around the world. Syria is just one of many regions where broken infrastructure can make supplies impossible to deliver by land. Drones are already busy monitoring poachers and providing aerial intelligence in disaster situations.

But Jacobsen didn't relate the Syria Airlift Project to any of those efforts. Instead, he looked back much further, to the Cold War when the Western Allies airlifted supplies into West Berlin. U.S. Air Force pilot Gail Halvorsen started a movement when he began dropping candy attached to handkerchief parachutes for children. Like the Candy Bomber, as Halvorsen became known, the drones could drop symbols of hope and happiness.

"People inside Syria affiliate airplanes with death. There are no positive memories of an airplane anymore," Ghosh-Siminoff says. "It would be nice to see a positive example of when a plane came to help them instead of to kill them. It would make them feel like they're not alone, that the world didn't forget them, and that there's still someone out there trying to help them."

OPEN SOURCE FILLS A GAP

With all the parts prepped, the group clusters around the launcher on the lake bed. A Waliid drone sits on top of two metal rails that will guide it out and up while the bungee accelerates it forward.

The final verbal checks ring out while a small crowd forms to watch.

"Clear!" Jacobsen shouts.

The launcher releases and the drone springs forward. The bungee falls away as the vehicle coasts upward and begins flying rectangles over Lake Lagunita.

Jacobsen runs a custom program that measures the plane's energy consumption at different flying speeds. Every so often, the Waliid increases its speed by 2mph, gradually moving from 28 to 50mph to help them determine what speed would give them the biggest flying range.

If Uplift begins sending drones into Syria, it will run another custom program. An app called Swarmify can take a single flight plan and turn it into as many semi-randomized flight paths as the team needs.

"Because every flight plan is slightly different, it ensures planes don't collide with each other," Jacobsen says. "It also gives you tactical survivability, because no two planes cross the same point on the ground. If somebody sees the first plane fly over, they won't catch the next one."

Much of the drone itself is made from off-the-shelf and open-source components. While Uplift could someday manufacture its own drones, right now it works with inexpensive hobby kits — the Waliid is actually the $100 Talon kit made by X-UAV. Its autopilot system is built by 3D Robotics. Its motors, props, and servos were all picked for their modest price, and can be found on Hobby King. This choice has its roots in the organization's origins, when Jacobsen had to teach himself the basics of building and programming a drone and fund the project inexpensively.

But even as Uplift's volunteer ranks grew, it kept building its own drones. It turns out that there isn't much of an alternative.

"When we looked at different airframes, one thing became immediately clear: The market is really polarized as far as cost and capability is concerned," Fetroe says. "If you tried to put all the planes in a line, and had the tiniest, cheapest one on one end, and some huge commercial or military drone on the other end, you notice there's a really big gap in the middle, kind of where we are trying to operate."

Drones that can carry more than a few pounds of cargo for an hour and cost, say, $1,000, didn't exist. Fetroe said new options are emerging, but most have yet to officially hit the market. For now, Uplift will carry on with its own design, which cost between $500 and $1,000 to build.

In its belly sits the real value — the payload. A wooden box, laser cut by Fetroe, opens to release its cargo. It floats to Earth strapped to a parachute made from garbage bags, or whatever other cheap plastic is available.

SYRIA IS NOT THE FINISH LINE

Whether or not the Syria Airlift Project succeeds, Uplift sees a future for its drones. What will start with just a few flights could scale to hundreds or thousands of planes that can feed entire neighborhoods. Even just a handful of planes can make rural

medical deliveries and bring aid to disaster-stricken regions where the political situation is more welcoming.

"My long-term goal is to help build a world where the use of starvation and medical deprivation are impossible — they just don't work anymore; you can always find a way to get humanitarian aid through. That's a lifelong ambition," Jacobsen said. "If we can get the first steps done, we can scale from there." ◐

Uplift Aeronautics members Heather Kline, Tomoki Eto, Mark Jacobsen, and Michael Taylor

A Guide to Drones
FROM THE SYRIA AIRLIFT PROJECT

THE NOUSHA
Easy hand launching makes this aircraft perfect for pilots to learn on.

THE ANSLEY PEACE DRONE
The platform for this drone allows room for experimentation with low-cost materials and mass production.

THE ISRAA
This aircraft has a low-cost design that is aerodynamic and easy to use.

THE WALIID
Its large volume and weight capability makes this aircraft great for short-range cargo delivery.

If you want to modify your own styrofoam aircraft to carry and release cargo, read this tutorial, penned by Uplift Aeronautics themselves, at makezine.com/go/waliid-cargo-dropper.

Written by Dale Dougherty · *Maker Pro Q&A*

Can Makers Survive Prime Time?

Intel CEO Brian Krzanich launches a build-off reality TV show

OVER THE PAST COUPLE OF YEARS, INTEL HAS BEEN MAKING ITSELF KNOWN IN THE MAKER COMMUNITY, releasing products like the Edison single-board computer, and contests aimed at the DIY electronics market. During his keynote at the Intel Developers Forum in San Francisco, Intel CEO Brian Krzanich announced a new TV show, *America's Greatest Makers*, produced by Mark Burnett (*Shark Tank, Survivor, The Apprentice*). The show will air in the spring of 2016, but first, 32 makers will be selected to compete for fame and a $1 million prize. Applications must be submitted by October 2 on the americasgreatestmaker.com website.

After last year's successful "Make It Wearable" competition, you must have decided to go bigger. Why?

Actually, the idea for a TV show was spawned by my wife and daughters. One of their favorite shows to watch is *Shark Tank*. When I came home from last year's competition, they said our wearables event was as interesting as *Shark Tank*. "Why don't you see if you could take the competition to TV," they suggested.

It started a yearlong effort for us. We wanted everybody to see how the whole making process works, how to build something that becomes a real product. This isn't something to be afraid of. We hope to get people from all different backgrounds, all different levels of capability and show that they are all able to come together and build a product.

What did you learn from last year's competition?

We learned a lot on the product side. It was a bit more difficult than we wanted for people to build with Edison. We are targeting Curie [Intel's new button-sized compute module with the Quark System-on-Chip, Bluetooth radio, and a six-axis sensor] for this competition, and we're really making sure all the software is ready. This platform will be much more robust.

We've improved Edison quite a bit between when it first came out and now. We've tried to take all that learning and carry it forward into Curie.

Makers won't know much about Curie before submitting their concepts.

I don't think in six weeks anybody could build something for the competition. We've asked people to submit a concept in written form, along with a video. That concept could be mocked up in paper or cardboard or whatever. The whole idea is to describe what they want to build and bring to market. Then we'll pick the 32 best.

You yourself are a Maker. People who work with you tell me that you talk about more than electronics, that you enjoy woodworking and welding. But this isn't just about your own interest. There's got to be a good strategic fit for Intel.

Well, there is. We want Intel to be an invention hub, whether you're building a server to power a big data application or the most low-end device with Curie. However, if we were to try to predict what people would do with new technology, we'd miss out on maybe half, or more, of the best ideas. So this kind of competition can help us see where things could go and what people want to have made. We can learn a lot. It pushes us into new areas and gets us into new partnerships.

For example, we originally built the RealSense camera so that you could step away from your PC and control the screen. Nobody who created RealSense thought we'd be using it on the top of drones to fly them autonomously through a forest. It wasn't until a bunch of people were goofing off, and they started saying, "I'd sure like to be able to fly this drone in follow-me mode and not have to worry about all the trees while I mountain bike." So it's not until you create things and unleash them do you see what the possibilities are. ◕

For more Maker Pro news and interviews, visit makezine.com/category/maker-pro, and subscribe to the Maker Pro Newsletter at makezine.com/maker-pro-newsletter.

DALE DOUGHERTY is the founder and Executive Chairman of Maker Media.

A CIRCUIT BOARD FACTORY ON YOUR DESK

BOTFACTORY'S INTEGRATED DESKTOP CIRCUIT BUILDER, **SQUINK**, IS A RAPID PROTOTYPING POWERHOUSE Written by Donald Bell

WHAT IF YOU COULD QUICKLY BUILD AN ENTIRE, WORKING CIRCUIT BOARD FROM START TO FINISH, without ever touching a soldering iron? That's the idea behind Squink, a $2,999 integrated desktop circuit fabricator from Brooklyn-based BotFactory. Their unique design helped the startup win the 2015 MakerCon Launch Pad competition in San Francisco.

Circuit printers are poised to be the next big thing in desktop fabrication. Many of the systems hitting the market are capable of creating conductive circuit traces, and even laying down solder paste for surface mount components.

Squink takes those board production functions and brings them a step further by integrating a pick-and-place feature that automatically populates both printed and professionally made circuit boards with all the miniature resistors, capacitors, and other components that you've specified. To accomplish this, it uses a vacuum pump with an integrated camera to pick up components from a tray, orient them correctly, and place them precisely where they are needed.

"Pick and place is the hardest thing to do in circuit manufacturing," says BotFactory's co-founder and CEO, Nicolas Vansnick. "We spent around 18 months developing our own pick-and-place software. Open software packages were too big and clunky. We got our own software down to around 25MB, which is really small and powerful. We can pick-and-place components in 10–15 seconds now."

BotFactory even uses the Squink to assemble circuit boards for other Squinks — their microfactory is already saving them time and money by keeping production in-house. And perhaps that's the best endorsement of all.

botfactory.co
makezine.com/go/botfactory-wins-launch-pad/

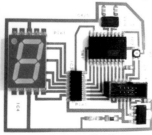

BotFactory

SPACE CITIZEN

THROUGH THE CENTURIES, WE'VE BEEN RULED BY A LOVE OF THE STARS.

Civilizations were built around their movement and meaning, and science is built around their discovery, measurement, and the pursuit of traveling to them. History is bookmarked with seminal moments in astronomy, each renewing our deep desire to climb ever higher in the sky. Falling apples and reflecting telescopes have led to footsteps on the moon, robots on Mars, and probes rushing ever farther past our solar system. Even more ambitious journeys will launch soon.

With the technological foundation firmly formed by giant space agencies and research facilities, powerful tools capable of major discoveries can now be built and accessed by at-home astronomers, garage spacesuit builders, and independent satellite designers everywhere. In this section we celebrate these amateurs and their increasing role in discovering our cosmos, and guide you to ways you can get involved. Take the stars, for they are ours. We are the Space Generation. ◗

Illustration by Nate Van Dyke

WRITTEN BY BRANDON LAWLER

REACH FOR THE STARS

I built my own Dobsonian telescope to see the cosmos up close.
Here's what I learned along the way.

BRANDON LAWLER is a hobbyist telescope maker and an active member of the Central Texas Astronomical Society. His passion is using CAD design and his homemade CNC router to bring large wood projects to life.

DOBSONIAN TELESCOPES ARE POPULAR WITH AMATEUR TELESCOPE MAKERS FOR THEIR EASE OF DESIGN AND CONSTRUCTION, PORTABILITY, AND THEIR USE OF LARGE OPTICAL MIRRORS. Pioneered by John Dobson in the 1960s, the instrument combines a Newtonian reflector telescope with a unique two-axis movable base. It uses a primary mirror to capture and reflect light, a secondary mirror to direct light into an eyepiece, and a focuser to make fine adjustments for viewing. The telescope's size is classified by the size of its mirror.

After picking up a copy of *The Dobsonian Telescope* by David Kriege, I built my first telescope with a 12½"-diameter mirror, then a 12" lightweight scope. Once I'd built a CNC router, I embarked on my third telescope, featuring a 16" primary mirror with aluminum trusses, wide vertical bearing arcs, a steel front-adjustable mirror cell, and a rotating base. The project took several months off and on to complete, although a skilled Maker could put a similar one together in a few weeks. I'm quite happy with the result, and the view in its large mirror is phenomenal.

All Dobsonian telescope projects are unique builds — here are the notes from my latest

THE RIG

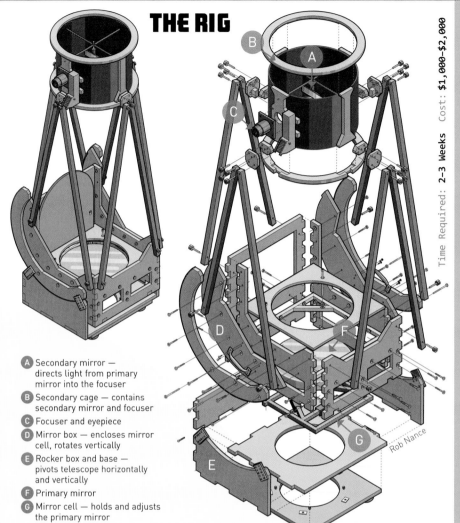

A. Secondary mirror — directs light from primary mirror into the focuser

B. Secondary cage — contains secondary mirror and focuser

C. Focuser and eyepiece

D. Mirror box — encloses mirror cell, rotates vertically

E. Rocker box and base — pivots telescope horizontally and vertically

F. Primary mirror

G. Mirror cell — holds and adjusts the primary mirror

Rob Nance

Time Required: 2–3 Weeks   Cost: $1,000–$2,000

Materials

» **Primary telescope mirror** aftermarket, homemade (see page 31), or custom-made. I found a used 16" primary on eBay.
» **Secondary mirror with "spider" holder** Search on eBay for a matched set. My secondary is 4"; for tips on choosing size, see garyseronik.com/?q=node/8.
» **Focuser**
» **Telrad finder scope**
» **Steel tubing,** ½"–¾" square, 8'-12'
» **Plywood,** ¾", 4'×8'
» **Kydex sheet,** black, 1/16" thick, 4'×4'
» **ABS plastic,** textured, ⅛" thick, 2'×4'
» **Teflon sheet,** ⅛" thick, 1" squares (8)
» **T-nuts,** ¼" with matching 2½" bolts and knobs (24)
» **Aluminum tubing,** thin wall, 1" square, 4' lengths (8) I got mine from Cycle 24.
» **Skate bearings (8)**
» **Flat braces,** 1½"×2½" (4)
» **Wood screws**
» **Wood glue**
» **Sanding and finishing materials**
» **Black woven velveteen,** 6'×8' Cut it so it stretches along the 6' axis.
» **Elastic cord, 4'**
» **Staples or finishing nails**

Tools

» **CAD software and CNC router (optional)**
» **CNC files (optional)** free download from makezine.com/go/dobsonian
» **Table saw and band saw**
» **Drill press**
» **Welder**
» **Screwdriver**
» **Sewing machine**
» **Scissors**

ESA/Hubble · Akira Fujii, Carrie Fay Lawler & Hep Svadja

Brandon Lawler

version to help get you familiar with the process and determine how you'll design yours.

1. CONSTRUCT THE MIRROR CELL

The core of the telescope, the steel mirror cell holds and adjusts the heavy, curved primary mirror. I welded mine from ¾" steel square tubing.

Because mirror flexure can distort an image, supporting the mirror properly involves building a "flotation" cell. The back of the mirror "floats" on 3 or more support points (this build uses 6) that are calculated using a software tool called PLOP (davidlewis toronto.com/plop). Given any mirror measurements, PLOP will provide the ideal support layout and how much distortion to expect for any number of flotation points.

As the telescope tilts toward the horizon, the mirror must be supported on its edge. Use the external Mirror Edge Support Calculator (cruxis.com/scope/mirroredge

calculator.htm) to decide whether to use a two-point, four-point ("whiffletree"), or sling support. While a sling or whiffletree provide the best edge support, a two-point edge support is much easier to construct.

The mirror itself must also be able to tilt in three dimensions in order to aim its light at the secondary mirror (a process called "collimation"). To do this, the mirror cell needs to be supported by 3 large bolts, at least 2 of which are adjustable. By adjusting the bolts, the mirror can be pointed toward the correct spot.

2. BUILD THE SECONDARY CAGE

This is the upper tube that contains the flat secondary mirror, Telrad finder, and focuser. In my build, the cage was cut on a CNC router from ¾" plywood, with threaded T-nuts added to support a truss assembly. The cage should be a hollow cylinder about ½" wider than the mirror, with the focuser

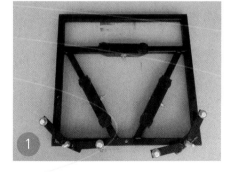

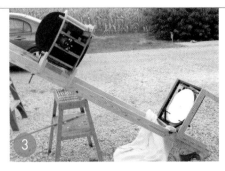

mounted directly facing the secondary mirror. The "spider," or secondary mirror holder, will suspend the mirror in the optical path in order to direct light into the focuser. Thin 1/16" Kydex plastic should be cut to length to line the inside of the cage as a "baffle" to block outside light.

3. MEASURE THE CELL-TO-CAGE SEPARATION

Every primary telescope mirror has a fixed focal length that's usually 4 to 6 times the width of the mirror. When you add the minimum distance from the eyepiece to the secondary mirror together with the distance from the secondary mirror to the primary, the total length should be the primary mirror's focal length.

In my build, the minimum eyepiece-to-secondary mirror distance (13½") plus the primary-to-secondary mirror distance (57½") equals the focal length of 71", which is roughly 4.5 times the width of the 16" mirror.

In order to check your measurements, you can construct jigs for your mirror cell and secondary cage, positioning them on a straight, adjustable track such as 2 planks of wood. Orient this assembly to allow you to view an object on the far horizon. Move the jigs until you can comfortably place a variety of eyepieces in the focuser and get a sharp image, then carefully measure the separation distances.

4. DESIGN THE MIRROR BOX

The mirror box encloses the mirror cell and allows the entire telescope to rotate vertically. I built mine entirely from CNC-cut ¾" plywood, and fastened it together with 2½" bolts. It features 2 semicircular arms, and mounting points for the trusses (T-nuts are fine), as well as a lid to keep the mirror safe when the telescope is not in use.

Building the mirror box is tricky, because the entire optical assembly (mirror cell, mirror box, trusses, secondary cage) must balance at the center of rotation of the arms. If the telescope rotates forward or backward on its own, then the mirror box is too deep or too shallow. Plan ahead by carefully weighing all components and approximating the center mass of the optical assembly.

Once built, line the undersides of the arms with textured ABS plastic as a bearing surface. Staples or finishing nails work fine, but be sure they don't bump up above the surface of the plastic. The plastic will ride on teflon pads, creating just enough traction for the telescope to avoid sliding on its own while not making it too difficult to point at things in the sky.

5. CUT THE TRUSSES

Thin-wall aluminum tubing is used to attach the mirror box to the secondary cage. While round tubing is sturdy, square tubing is easier to work with. Once it's cut to length, drill a hole through each end of the tubing with a drill press. During on-site assembly, attach each truss to its mounting T-nut using a bolt with a thumbscrew knob.

Using plywood, make 4 attachment blocks to pair the trusses together and create a small ledge for the secondary cage to rest on while you secure it. During assembly, you'll attach the trusses to the mirror box, then sit the secondary cage on top and bolt everything in place.

6. BUILD THE ROCKER BOX AND BASE

Made of ¾" plywood, the rocker box supports the mirror box on 1" teflon pads, allowing it to rotate vertically. The rocker arms must also have guides to keep it on the track; flat metal 1½"×2½" braces lined with ABS plastic work nicely. The box should be deep enough to allow the mirror box to swing all the way down. Line the underside of the rocker box with a ring of ABS plastic to allow it to ride on the base.

The base of the telescope should be a wide, sturdy square or circle of wood with teflon bearing pads matched to the ABS plastic ring of the rocker box. And the legs of the base should be as wide as possible to accommodate weight imbalance as the telescope is moved around, to avoid tipping. The base and the rocker box in my telescope are secured with a skate bearing assembly but could be more simply attached with a bolt through the center.

7. FINISH WITH A LIGHT-BLOCKING SHROUD

Once your telescope is assembled, drape black woven velveteen around the truss assembly, clipping it with safety pins. Make sure the shroud can slip on and off of the

Carrie Fay Lawler

assembled scope, and that it stretches along the circumference (not lengthwise). Sew the seam, and sew an elastic cord into the top to secure it to the secondary cage. Trim off any unused fabric along the base.

• • •

Dobsonians come in all shapes and sizes, from small 4"-8" builds all the way up to massive 24"-36" creations. Because the basic movement and optics requirements are relatively simple, much of the design is left to the builder. Be creative! And, once your telescope is finished, be sure to join a local astronomy club to learn more about the sky and share your hobby with others. ◐

MIRROR GRINDING

WRITTEN BY DONALD BELL

Aside from being more cost-effective, handmade mirrors are prized for their superior optics compared to machine-made mirrors.

It is, however, a time-intensive project. An 8" mirror requires shaping and polishing a glass blank by hand for around 40 hours using a grinding block and progressively finer grits of grinding and polishing compound. Larger mirrors take considerably more time to create, but also represent the best value, as large, high-quality, commercially made telescope mirrors can be prohibitively expensive.

If you're curious to give it a try, it's recommended to start with a small mirror (8" or less) for faster results and to understand the process fully before embarking on a larger mirror. Check out local astronomy and telescope clubs in your area to see if they offer any mirror-making workshops.

Get the CNC files and more details at makezine.com/go/dobsonian.

BARN DOOR SKY TRACKERS

WRITTEN BY KEITH HAMMOND

To photograph the stars, you need a gadget that can track the revolving night sky in a perfectly timed arc. Otherwise all you'll see is streaks and blurs.

You can buy fancy motorized "equatorial mounts" for telescopes and cameras, but it's way cheaper and more satisfying to build your own simple "barn door" tracking mount using a long bolt or threaded rod as a drive screw. You mount your camera on the "door," then aim the hinge straight at the North Star, Polaris. The motor opens the door very slowly to match the sky's rotation, for blur-free exposures of minutes or even hours. You can set the speed using a microcontroller or a simple circuit.

But there's a catch: A straight drive screw turned at a constant rate won't produce a constant angular motion. It's called the "tangent error" — and here's how some of our favorite DIY barn door trackers solved it.

ANALOG

Sky & Telescope contributing editor Gary Seronik of Victoria, British Columbia, built a lightweight, portable tracker that drives a simple 4RPM **DC motor** with an **adjustable voltage regulator** to dial the rotation rate, and a **curved bolt** to reduce tangent error (Figure Ⓐ). He's shared his design and schematics in a great tutorial at garyseronik.com/?q=node/52. "It's hard to beat a DC motor and simple regulator circuit for simplicity and performance," he says.

Seronik also created a smaller Hinge Tracker (garyseronik.com/?q=node/184) using an **8" strap hinge** in place of the plywood. A straight bolt introduces tangent error, but he solves that by taking shorter exposures and "stacking" them in freeware called DeepSky Stacker.

» Visit *Make:* online for our exclusive how-to on building a motorized Hinge Tracker, at **makezine.com/go/hinge-tracker**.

DIGITAL

Chris Peterson in Guffey, Colorado, used a **straight bolt in pivoting mounts** and cleverly programmed a Freescale/Motorola **68HC705C8 microcontroller**

to drive a **1.8° stepper motor** at a **variable rate** to produce constant angular motion. He's taking 20-minute exposures with a 300mm lens, and has shared his schematics and code at cloudbait. com/projects.

University student David Hash (now an aerospace engineer) updated Peterson's build with an **Arduino Pro microcontroller, 1.8° stepper, and Pololu microstepping driver board** to give 3,200 microsteps per rotation (Figure Ⓑ). He's shared his build and code on Reddit at makezine.com/go/hash-barn-door. He gets great photos by stacking multiple 90-second exposures; check out his Andromeda Galaxy pictured above, and more at imgur.com/a/GbMaj#0.

Finally, Alex Kuzmuk in the Ukraine designed **custom laser-cut doors and acrylic gears** for his curved-bolt tracker, and drove it with a **$2 stepper, Arduino Uno, and LCD Keypad Shield** for easy speed adjustment. Check out his build and code at makezine.com/go/kuzmuk-barn-door. ◐

David Hash and Gary Seronik

SOLAR SAILOR

WRITTEN BY NATHAN HURST

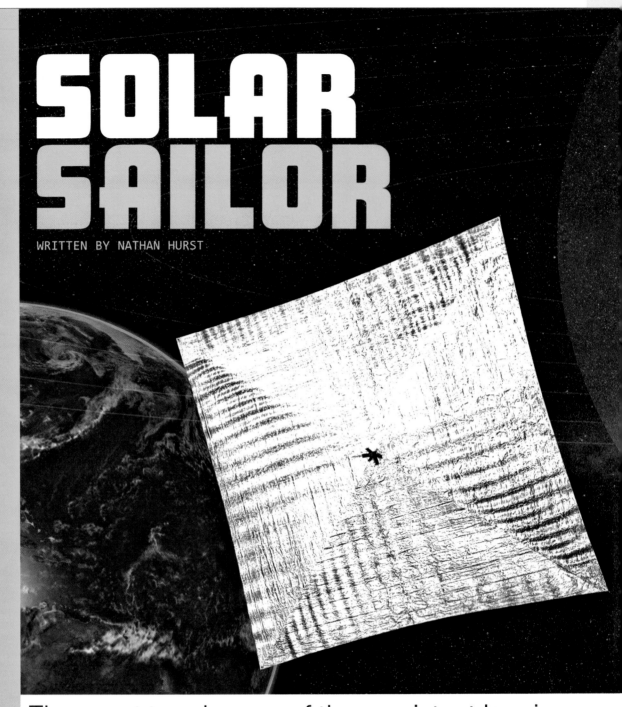

The quest to solve one of the persistent barriers to citizen space exploration

NATHAN HURST is an editor with *Make*. When he's not writing or editing stories, he can be found riding a bicycle he built and going on camping treks, often at the same time.

ON JUNE 7, IN THE SKY SOMEWHERE OVER BAJA CALIFORNIA, MEXICO, a small satellite opened and extended a wide, thin mylar membrane, like a curious flower blooming, or a space-soaring butterfly emerging from its cocoon. Launched on an Atlas V, a two-stage, 190-foot rocket with expandable payload, the LightSail had floated, dormant, in free fall in an orbit between 350 and 700km around the planet for 18 days. It was a culmination of decades of effort, and a concerned crew of scientists and observers worked out bugs and tracked its location.

Bill Nye — he of Science Guy fame — was quick to tweet: "Sail has Deployed!" Though

the effort to launch a solar sail pre-existed Nye's tenure as CEO of The Planetary Society, it was thanks in part to him that LightSail reached the sky — and the hearts and minds of a new generation of space enthusiasts. It represents, according to project manager Doug Stetson, an important step: A compact, cheap, indefinite supply of propulsion for space vehicles — and not just NASA-sized spacecraft.

"Democratization of space is really the ability to be able to send those small exploration spacecraft beyond Earth, which is something that's been, really, not possible before," Stetson says.

With solar sailing technology, Stetson explains, it'll be possible to send small (cheap) spacecraft beyond Earth orbit. "One could envision a day when small organizations, universities, student groups, nonprofits, things like that could actually have their own interplanetary exploration spacecraft to send to the moon, or near earth asteroid, or even Mars."

DO IT YOURSELF SPACE PROBES
Nye, Stetson, and The Planetary Society leveraged Kickstarter, as well as modern

tech like CubeSats and microcontrollers, to make LightSail happen. The 18.4'×18.4' sail packed down to the size of a loaf of bread to fit in the standardized, 3U-size CubeSat case, and deployed in a controlled manner so as not to rip the thin, 4.5µ mylar apart.

Once unfurled, the LightSail remained in orbit for barely a week before burning up on reentry. After all, it was just a trial run to test deployment. At its highest — still in low-earth orbit — the sail was still subject to too much drag, thanks to the atmosphere, to take advantage of the solar push. Next year, The Planetary Society plans to launch a similar sail, but to nearly 800,000 feet — more than half the height of the ISS — where it'll actually be able to sail on solar wind. As it orbits Earth, crew on the ground will adjust its orientation, using a momentum wheel to make it tack across the solar photon stream, or run along with it. If it goes well, the next version will stay active for up to six months, eventually reentering once its motion is no longer controlled.

THE PROPULSION PROBLEM
The LightSail, both this year's launch and next year's controllable flight, represents

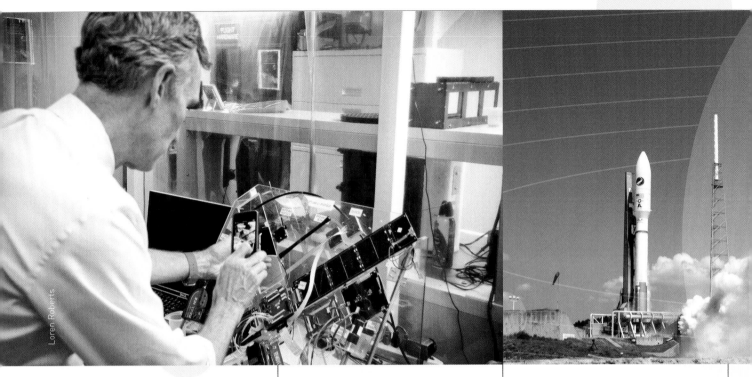

Loren Roberts

a particular advancement for anyone who wants to run their own experiments in space. Between CubeSats and the miniaturization of control technology (see "Space Chase," *Make:* Volume 46), it's easier than ever to get your experiments into orbit. (Still not easy, many would point out. Just easi*er*.) What hasn't gotten easier is moving them around once they're there.

"The real limitation has been propulsion, because typically, propulsion is very heavy," says Stetson. "And that's where LightSail comes in. It's the chance to put inside of a CubeSat its own intrinsic propulsion."

Traditional propulsion has been chemical — that is, fuel based — and is still necessary to lift LightSail and other solar sails to space. Though it does provide quick acceleration, fuel is heavy and gets used up. With a solar sail, the acceleration comes from photons — which have energy and momentum — striking a reflective surface and transferring tiny bits of momentum to it. The sail accelerates slowly, but indefinitely, making it ideal for extreme long distances and lightweight, low-cost missions.

In fact, the LightSail is not a new concept. Quite the opposite: As space propulsion goes, solar sailing is one of the most venerable goals of astronomers and engineers, dating back as far as Galileo. A new generation's love affair with space shows the cyclical curve of history wending back

to the era of our parents, when the U.S., the USSR, and NASA threw money at the sky and anything was possible. It still is — it just costs a lot less. For LightSail, The Planetary Society estimates around $5.5 million, raised mostly through private donations and grants, and offset by a million-dollar-plus Kickstarter that coincided with the trial run. (Low cost is a relative term when we're talking about space.)

"It's always been the case that until a technology like that is demonstrated in orbit, it won't reach the mainstream," says Stetson. "Organizations like NASA will not be willing to commit large amounts of money to missions in the solar system until those key technologies are proven."

That's a gap The Planetary Society intends to fill, and the ultimate goal for LightSail. The 2016 edition won't go interplanetary, nor will it have a particular mission. It will act as proof of concept, relaying information about how the sail moves when photons are acting upon it.

A COSMIC FIX

The 25 days LightSail spent in orbit were filled with a series of triumphs and pitfalls. Remember that controlled deploy? It ran on common hardware. Designer Chris Biddy used brushless DC motors with a planetary gear set, driven by a PIC controller programmed by Alex Diaz. Hall effect sensors

tracked the deployment so the operators could know how far it had extended, and a worm gear held it in place during idle times.

But for a controlled deploy to work, you need to be able to control it. And one of the first things that happened was LightSail's communication went dark. It froze up while sending data packets home — the satellite equivalent of the blue screen of death or the dreaded spinning beach ball. Engineers on the ground can't exactly hold down the power button, but each time the satellite passed over Cal Poly or Georgia Tech — the two schools partnered with The Planetary Society — the team sent a reboot command.

Eight days later, LightSail began sending data again, although it seems that wasn't because of earthbound attempts. Instead, the reboot was likely caused by a stray cosmic ray, which strike the satellite's electronics with surprising regularity.

Operating on the "photos or it didn't happen" principle, LightSail began sending garbled jpegs home, a bit at a time, each time it passed Cal Poly or Georgia Tech. Then, after the external panels swung open, it delivered a low-battery signal and went silent again.

"It was a roller coaster for sure," says Biddy. "You go into the weekend thinking, man, it didn't make it. Such a bummer. And then I got a text or something from one of the operators saying, hey, it's back."

Navid Baraty

Josh Spradling

Josh Spradling

RISING AGAIN

Like the journey of LightSail itself, but
expanded over decades rather than days,
The Planetary Society's trajectory is a long
slog full of struggles, fails, and false starts.
Founded in 1980 by Carl Sagan, Louis
Friedman, and Bruce Murray because
NASA scrapped plans for its own solar
sail, the society relies on private funding
to pursue its exploratory space missions.
(It's also the organization responsible for
the Search for Extraterrestrial Intelligence,
or SETI, and Planetary Defense asteroid
tracking.) So when funding fell through or
launches got postponed, or even exploded
after liftoff — a 2005 solar sail was lost
when the Cosmos 1 rocket it was on
exploded — the project sat, not so much
gathering dust as awaiting technological
and sociological advances, and funds.

Kickstarter, CubeSats, and the Science
Guy provided the oomph. And in space,
LightSail's battery stabilized, and the team
initiated deployment. Over the next few days,
the satellite gradually beamed back photos
indicating that it had indeed stretched
out. Then, a few days later its orbit fully
deteriorated and LightSail disintegrated.
Time of death: 1:23 pm EDT, June 15, 2015.
LightSail Two, time of launch: Coming
September 2016. ◗

WRITTEN BY DAN RASMUSSEN

NEAR-SPACE PHOTOGRAPHY
with APRS Radio Tracking

Dan and Emma Rasmussen

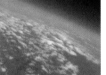

Track a hacked Canon camera to the stratosphere and back, to photograph the blackness of space

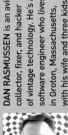

DAN RASMUSSEN is an avid collector, fixer, and hacker of vintage technology. He's a software engineer who lives in Groton, Massachusetts, with his wife and three kids.

AFTER READING THE FIRST DIY SPACE ISSUE OF *MAKE:* **(VOLUME 24),** my daughter Emma and I were inspired to work on a near-space balloon to photograph the stratosphere. At the time, the standard recovery method was a cellphone programmed to communicate its position after landing. Unfortunately, once you let your balloon go, you had no idea where it was until it burst, parachuted back down to earth, and finally phoned home. We wanted a way to get live reports from our flying machine while it was in flight.

The solution is the Automatic Packet Reporting System, an amateur radio network that reports the location and other data about the sender. Package up a radio transmitter with a GPS receiver and some sensors in an Arduino shield, program it to send APRS packets, and you have the Trackuino — an open source tracker you can use to locate anything (provided you've got your ham radio license).

Then add a Canon camera hacked with

THE RIG

1. Weather balloon
2. Parachute
3. Duct-taped foam capsule
4. Buzzer
5. Hand warmer packet
6. GPS antenna
7. Lithium AA battery pack
8. Canon point-and-shoot camera, hacked with CDHK firmware upgrade

Time Required: 2–3 Weekends Cost: $550–$600

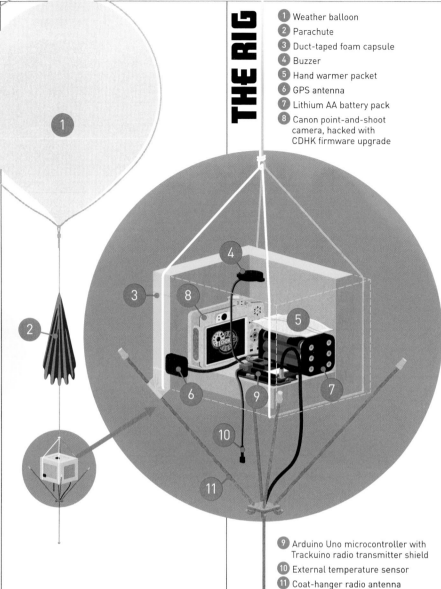

9. Arduino Uno microcontroller with Trackuino radio transmitter shield
10. External temperature sensor
11. Coat-hanger radio antenna

James Provost

CHDK firmware that lets you automate your photography, and you have a tried-and-true and cost-effective recipe for a thrilling, live-action, near-space adventure.

Emma and I launched our balloon from North Adams, Massachusetts. Within about 2 hours it would climb to 94,000 feet, attain ground speeds greater than 100mph, and record hundreds of stunning photos and videos. All the while telling us exactly where it was and what was going on.

Here's an overview of how to build your own Trackuino-based high-altitude balloon rig for photographing the stratosphere. For complete step-by-step instructions, visit the project page at makezine.com/go/near-space-balloon-cam.

AUTOMATIC PACKET REPORTING SYSTEM (APRS)

APRS is an amateur radio protocol designed to track a sending station that's on the move. The fantastic thing about APRS is once you broadcast the packets, everything else is done. Thanks to ham radio operators around the world, there are receiving stations listening for APRS packets nearly anywhere you might be. These stations forward your packet to internet sites that provide near real-time reports of your location, altitude, speed, and any other data you can fit into an APRS packet. Just surf to aprs.fi, enter your ham radio call sign, and you'll see an interactive map of your flight with all this information. Anyone can track

Materials

VISIT MAKEZINE.COM/GO/NEAR-SPACE-BALLOON-CAM FOR MORE VENDORS AND PART NUMBERS.

- » **Arduino Uno microcontroller board** Maker Shed #MKSP11, makershed.com
- » **Canon digital camera, CHDK compatible, with 10GB memory card** We used an A560.
- » **Trackuino PC board, Version 2.2** from oshpark.com or other online PCB houses
- » **Radio transmitter, VHF narrow band FM, high power** Lemos Int'l #HX1-144.390-3, lemosint.com. For U.S. amateur frequency (144.39MHz), buyers must provide their amateur radio call sign when ordering.
- » **SparkFun Venus GPS with SMA connector**
- » **Resistors, ¼W, 1%, 1206 SMD package:** 6.81kΩ (1), 2.21kΩ (1), and 1kΩ (1)
- » **Switch, tactile, SPST, normally open, 0.05A 12V**
- » **Terminal blocks, PCB mount, 5mm:** 3-position (1) and 2-position (2)
- » **Temperature sensors, 2.7V, TO92-3 (2)**
- » **Noninverting buffer ICs, CMOS logic level shifter, 5TSOP SMD package (3)**
- » **Capacitors, tantalum, 10µF, 16V, 10% (2)**
- » **MOSFET, N-channel, 30V 4.8A, SOT23 SMD package**
- » **SMA connector, female (jack), 50Ω**
- » **Voltage regulator, LDO type, 3.3V 1A, SOT223-3 SMD package**
- » **Buzzer, loud**
- » **Cable ties, various**
- » **Helium** about 100 cubic feet for 2lb payload
- » **Balloon, 600g** We used a Kaymont.
- » **Parachute, RocketMan 4 foot**
- » **Parachute cord**
- » **Coat hangers or piano wire, 8'** for antenna
- » **Coax connector, UHF female to SO239 female, 4-hole panel mount, with solder cup** for antenna
- » **Foam insulation board** from a hardware store
- » **Gorilla Glue**
- » **Duct tape, high visibility**
- » **PVC pipe, 1", 12" length** for fill rig
- » **PVC fitting, elbow** to fit brass adapter
- » **Brass pipe fitting adapter** to mate your PVC elbow to your helium regulator hose
- » **Teflon tape**
- » **Batteries, lithium, AA size (8)**
- » **Battery clip, 9V, with 5.5mm/2.1mm plug** to connect the battery pack to the Arduino
- » **Battery holder, 6xAA**
- » **Hookup wire** Maker Shed #MKEE3
- » **Hand warmer packet**

Tools

- » **Computer running Arduino IDE** free from arduino.cc/downloads
- » **Radio receiver, portable** covering APRS frequency for your region (North America is 144.39MHz). We used an older BCT7 Uniden scanner with a portable power supply.
- » **Android device with APRSdroid app**
- » **Audio patch cord**
- » **Soldering iron, temperature controlled with solder and SMD-compatible tip**
- » **Dental pick** or similar tool, for handling tiny parts. Harbor Freight sells an inexpensive kit.
- » **Voltmeter** auto-range recommended
- » **PCB vise (optional)** makes life easier
- » **Compressor hose** to fit your fill rig and the regulator on your helium bottle
- » **Dummy load** for testing only, 1W–2W rating
- » **Wire cutters or flush snips**
- » **Knife or saw**

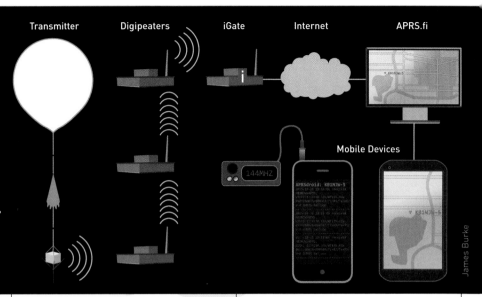

HOW APRS TRACKING WORKS

The APRS amateur radio network passes along digital data packets sent by your transmitter, so you can track its location.

» Local repeater stations ("digipeaters") re-transmit your packets, increasing the likelihood they'll be picked up by a station connected to the internet.

» Internet-gateway stations ("iGates") forward your packets to the APRS Internet Service (APRS-IS) network. Websites can then plot your transmitter's location on a Google map, viewable from any browser.

» In the field, you can receive packets directly on your own radio receiver and decode them with an Android app.

Transmitter Digipeaters iGate Internet APRS.fi

Mobile Devices

James Burke

your flight live and even help with the chase.

You can also directly receive packets on a radio or scanner capable of tuning the APRS frequency (144.39MHz in North America). This is important after landing — chances are, the final packet that reaches aprs.fi will get you close, but you'll need to find the resting location yourself.

APRSdroid is an Android app that decodes APRS packets, which are encoded to analog using Audio Frequency Shift Keying (AFSK). Just tune your receiver to 144.39MHz and use an audio patch cord to plumb it into your Android device and decode your packets.

THE TRACKUINO

The Trackuino (Figure Ⓐ) is an Arduino shield that combines a low-power radio transmitter with a GPS receiver and other sensors to let you track your balloon via APRS, including GPS location, altitude, speed, internal and external temperature, and battery state. It can also drive a beeper or buzzer (to help you find which tree it's stuck in during the recovery phase of your adventure).

The heart of the Trackuino is its 300-milliwatt radio transmitter made by Radiometrix. It's controlled by the Trackuino firmware (github.com/trackuino/trackuino) to periodically broadcast packets on your local APRS frequency. On the ground, the transmitter isn't powerful enough to be heard by a receiving station unless you're right on top of it. Up in the air, though, wow — it can be heard for hundreds of miles.

The Trackuino isn't available pre-built —

you'll buy the circuit board and then solder the components by hand. If you're new to SMD components, read *Make:*'s soldering primer at makezine.com/go/solder-smd and consider buying an SMD practice kit.

NOTE: A new board called Tracksoar, available for pre-order now, is an evolution of the Trackuino that's more compact, lighter, and more efficient; watch for an upcoming review in *Make:*.

HAM RADIO LICENSE

Radio skills can seem old school to us smartphone-toting 21st-century citizens. In fact, Wi-Fi, Bluetooth, 3G and 4G networks, and, yes, your cellphone are all radios. Understanding and mastering radio communications is very much a 21st-century skill.

Besides allowing you to use APRS, getting your entry-level ham radio license (Technician) will require you to learn some basics of radio technology and electronics; stuff you probably already want to know. There are a lot of study guides out there; get started at the American Radio Relay League (arrl.org).

THE CAMERA

We weren't ready to send a $500 camera off into the ether when there was a reasonable chance we'd never see it again — we needed something economical that could take photos and videos on a timer. The solution is the Canon Hack Development Kit (chdk.wikia.com), a free firmware enhancement for Canon digital cameras. An inexpensive point-and-shoot (we found

an A560 (Figure Ⓑ) on eBay for $35) loaded with CHDK can be programmed to take photos and videos periodically — say, a photo every 5 seconds and a 30-second video every 5 minutes — just what you need to visually record your adventure.

POWER

You need at least a couple of hours of run time for the camera and much longer for the Trackuino (in case recovery takes longer than expected). The easy answer is lithium batteries — more power for the weight. The Canon A560 takes two AAs and was still running when we recovered it, hours after launch.

The Trackuino can operate at a range of voltages using four, six, or eight AA batteries. We did the math: Four provide insufficient run time; eight add too much weight. Our Trackuino ran more than 6 hours on six batteries, with plenty of life left when we recovered the package.

ANTENNA

Build a simple quarter-wave ground plane antenna from scavenged parts as described on the Trackuino site (makezine.com/go/trackuino-ant). It's very important to test and tune your antenna with an SWR meter (along with someone who knows how to use it). This is where your local ham club can be very helpful.

ENCLOSURE

Insulating foam board is inexpensive and easy to work with. Assemble a lightweight, sturdy capsule from it using Gorilla Glue and

Dan Rasmussen

high-visibility duct tape. Put a hand warmer inside — electronics don't like the cold.

HELIUM AND BALLOON

How much helium do you need? It depends on the size of your balloon, the weight of your payload, and how much free lift you want — the surplus lift after all the weight is countered by the helium. Too little free lift and you risk a "floater" (a balloon that never bursts), too much and your balloon bursts at a less exciting altitude. We used a 600g Kaymont balloon and had a 2.2-pound (1kg) payload; using the online CUSF Balloon Burst Calculator (habhub.org/calc) we determined about 2 pounds is a safe amount of free lift for a 5m/s ascent rate, so we'd need about 100 cubic feet of helium for a successful flight to about 90,000 feet.

You'll fill the balloon to your particular volume — to about 6 feet in diameter, in our case. Then, as it rises into the atmosphere, the pressure drops and the balloon expands. This particular balloon is rated for a burst diameter of 20 feet!

LAUNCH SITE, WEIGHT & THE FAA

Initially we planned on launching from our front yard in Groton, Massachusetts, but when we looked at balloon flight predictors online we found, more often than not, our balloon would wind up in the ocean. Finally we settled on North Adams as our launch site. The easiest predictor to use was Habhub's (habhub.org/predict) — and it was accurate with its prediction (Figure C).

Once you've chosen a safe launch site, call your local FAA field office, weeks in advance, and tell them what you're planning on doing. They'll want to know when and where you're launching, the size of your payload, your expected ascent/descent rate, and the expected path your balloon will follow. Keep your payload under 4 pounds — if you go over 4 pounds, the rules change drastically. To keep things inexpensive, aim for about 2 pounds.

If your launch site is near any small airports, you may need to file a NOTAM (Notice to Airmen) — our FAA representative helped walk us through this process as well. (I found that they were nearly as excited about our project as we were.)

Finally, use common sense. Scan the skies for nearby aircraft and wait for them to clear the area before launching.

PRACTICE, PRACTICE, PRACTICE

This is a complex system and lots can go wrong. You need your transmitter and GPS receiver working properly, the antenna properly tuned, the Trackuino program properly configured and running, the camera switched on with your CHDK program running, the balloon filled with the correct amount of helium, and everything sealed up in a near-space-ready, secure, warm enclosure.

Test your camera by setting it up in your front window before you leave for work. Test the Trackuino by taking it on long drives; have someone hide it in your neighborhood, then go find it using your receiver and APRSdroid. Engage with your local ham club — I guarantee they'll have good suggestions to offer, and maybe even some helpers. The more you practice, the more likely you'll succeed.

LAUNCH AND CHASE

On launch day, Emma and I woke up very early for our 2-hour drive to North Adams. We were set up by 8:45 a.m. and had everything assembled and launched at 9:30. It was hectic but, wow, what a moment. Letting it go was a highlight of the day (Figure D).

But that was just the beginning of the adventure. As soon as our balloon hit the sky, it was on aprs.fi. Soon the updates started rolling in: 20,000 feet, 40,000, then 80,000, 90,000 ... just when we were starting to worry about it becoming a floater, we learned it was finally descending after hitting 94,000 feet. All the while, we were getting advice on which way to head from our network of internet and radio club observers. Just as Habhub predicted, it changed direction on us a couple of times, but in the end, it landed right where it was supposed to.

RECOVERY

Recovery was its own adventure. With the help of local hams we tracked our payload to its landing site about 6 hours later. My thanks to all those who helped track it down, and to the chainsaw-wielding landowners, Fred and John, who got it down the last 30 feet (those two trees were destined to become firewood anyway). It was one of the most enjoyable and rewarding days of our lives. ●

HABHUB-PREDICTED TRACK

ACTUAL TRACK

Get complete instructions and code, stratosphere photos and videos, and share your high-altitude balloon (HAB) projects at makezine.com/go/near-space-balloon-cam.

MAKERS on MARS

Adam Savage geeks out with author **Andy Weir** about his upcoming MacGyver-in-space movie, *The Martian*.

filled with technical details on hacked-together and makeshift survival solutions, *The Martian* follows the story of astronaut Mark Watney's struggle to stay alive on Mars after being abandoned and left for dead by his crewmates. It explores practical applications of chemistry, electronics, physics, as well as the mindset of a never-say-die Maker/astronaut (played by Matt Damon in the upcoming cinematic release), all with clarity and humor.

This interview of best-selling author Andy Weir by MythBusters star Adam Savage was originally created for Tested.com's series The Talking Room. You can find the full interview, along with every episode of The Talking Room, at j.mp/talkingroom.

Adam Savage: Talk to me about the genesis of this book.
Andy Weir: Well, I'm kind of a dork, and I really like to sit around and think about the space agency and space program and manned and unmanned space-flight. I was thinking, how could we do a manned Mars mission with the technology we have right now? What would it entail, how do you get people to Mars, how do you get them to the surface, how do you get them back up? So I came up with what's basically the Ares mission as described in the book.

AS: So is the book the end result of an exercise in your head of "How will this be possible, and what would happen?"
AW: Yeah. I was basically saying, OK, how would we put together a

Tested.com

ANDY WEIR is the author of *The Martian*, which is being made into a major motion picture. As a self-proclaimed "lifelong space nerd," he is a devoted hobbyist of subjects like relativistic physics, orbital mechanics, and the history of manned spaceflight.

ADAM SAVAGE is the co-host of MythBusters. Previously, he worked in the special effects industry as an artist, fabricator, and modelmaker. He has spent his life gathering skills that allow him to take what's in his brain and make it real.

Mars mission the way I want to do it? [It is] a manned mission, so you can't just have a failure kill the crew. You have to say, what happens if this fails? How do you ensure the crew survives? How do you abort at this point in the mission?

AS: So in each of these things you're trying to think like NASA would think it terms of worst-case scenarios, and you're going through space history to see how NASA solved these types of problems in the past?

AW: Absolutely. The fun part is thinking of a problem that's never been solved, and what can I do to make sure it gets solved? What if they have to abort when they're halfway to Mars? What if they have some critical failure on their ship? Is there an orbital trajectory that just brings them back to Earth fairly quickly without ever going to Mars? So I was working on all that and I thought of all these things that could go wrong on the surface. This could break, that could break, these two things could break at the same time, how do we deal with that. And then I started to realize the increasingly desperate solutions, and I thought, "Well, theoretically they could survive if they did this and that," and I realized that makes for a pretty interesting story. So I created a poor hapless main character and subjected him to all of them.

AS: Is there a history of a certain mission from NASA that helped really inspire or guide your thinking?

AW: Obviously, Apollo 13. My favorite scene in the Apollo 13 movie is when they're making the CO_2 scrubber adapter.

20th Century Fox

> I would always start with the problem and say, "OK, how does he solve it." I wanted each problem to come from the solutions to the previous problem.

Basically the "mailbox" it was called. That's one of my favorite scenes of any kind in cinematic history. And so basically I wanted to make a whole book of that. That's what I was going for.

AS: Was there a point in which you were writing, thinking through what could go wrong and how would you fix it, where you went to a dead end? Where you went down a path that was just too difficult to fix and you backtracked?

AW: Yes, that actually happened a few times. I would always start with the problem and say, "OK, how does he solve it?" I wanted each problem to come from the solutions to the previous problem. The whole book I wanted to be kind of a cascade failure. I wanted each thing to lead to the next.

A SPACE DORK'S LIBRARY

AS: The thing that I find surprising is the way you cover the astronauts. I know that there's lots in the literature of NASA about their evolving understanding of the personality type that makes a good astronaut.

AW: The astronauts are just universally acknowledged as a cut above the rest of humanity. They're just so awesome, and you say, "What makes a good astronaut?"

If you're going to have a mission that's going to take an entire year from start to finish, you'd better have six people who work together. I usually don't like it when I'm watching a movie or reading a book and there is a crew, and they have this tension and these personal issues and stuff like that. I'm like, "No, astronauts are *unbelievably* professional."

So the things I wanted to do for the crew are, first off, obviously everybody is extremely good at their technical skills. Everybody knows exactly what they're doing. Number two, a very good esprit de corps. They have no personal issues, no arguments or problems with each other, they don't butt heads on anything. They're really professional and they get along very well. Then third, a deep and profound confidence in their commander. Commander Lewis, nobody ever questions what she has to say, ever, and not because they're afraid of her, but just because they respect her that much. They have complete faith in her. And you need that in a crew of anything from a fishing boat to a spacecraft.

AS: Is your reference library like 20 books or 100 books or 500 books?

AW: It's one computer. I just Google around to find what I want and then make sure I get good sources.

To be fair I should point out that [the mission is] about 95% Mars Direct by Robert Zubrin. His concept was to send a ship to Mars and then make fuel using the Martian atmosphere. It's called In-Situ Resource Utilization.

AS: OK, and that saves a tremendous amount of payload into getting stuff there.

AW: Right. That's roughly what Zubrin's idea was.

He came up with that before ion engines existed. [In *The Martian*, the spaceship that goes from Earth orbit to Mars orbit uses ion engines. The surface-to-space vehicles, however, use traditional rockets.] Ion engines are real technology. They're not just invented for the book. Basically they're particle accelerators that shoot argon out very, very fast, so fast that the particles gain relativistic mass.

AS: Oh, wow. This means that it accelerates very slowly, but consistently over a long period of time?

AW: A very long time, and you need a lot of energy to do that, so you need a reactor aboard. Then you accelerate the ship at two millimeters per second per second, so it would get going faster and faster. If you do that for months, you end up going very fast. You need a delta-v of two and half kilometers a second to put yourself on a Mars intercept, which is 2,500 meters per second, which is about 5,000 miles an hour.

DOWN TO THE DAY

AW: So I made this simulation, and naturally I had to have a launch window. I needed to know the locations of Earth and Mars. I had to calculate this trajectory where you have Earth moving, Mars moving, and then my ship trying to match speeds and location with Mars.

AS: So did you choose a year then?
AW: I had to choose a *specific date*.

AS: This is so great! This is exactly the kind of level of detail I go into with my projects.
AW: Well, because of plot reasons the originally scheduled mission has to overlap Thanksgiving. So I had to choose a launch date such that they will be there on Thanksgiving of that year, and Thanksgiving floats around, so that's kind of a pain in the ass.

FEEDBACK FROM THE PROS

AS: Let's go back to the first time a NASA engineer writes to you. It's probably before the book is published, right? Is there somebody who'd worked at NASA that sort of came across the book?
AW: Yeah. I mean I had people who are engineers, scientists, and people who work for NASA or JPL would send me email with either critiques or, "Hey, good job." The ones that I really liked are the ones where they sent "Hey, you got the math wrong," because then I feel really good about it because it's been double-checked.

There is one scene early on in the book where Watney is reducing hydrazine to liberate the hydrogen, and then he's burning the hydrogen with oxygen to turn it into water. This does not go as smoothly as he'd like ... but the point is he does end up reducing a certain amount of hydrazine and turning it into water over a certain period of time.

I got email from a chemist saying "OK, you tell me how

much hydrazine he changed and how much water he made, which is correct," and he said "but you've also in the past told me the atmospheric pressure of the habitat, and the general dimensions of the habitat, from which I can calculate the volume, and you told me how long it took him to reduce all these hydrazine. It's extremely exothermic to do that reaction and from that I can calculate ...." Basically he had enough information to calculate that temperature increase in the habitat that would happen as a result, and it'd go up like 400 kelvin, so Watney would have died. He would have like roasted himself alive. I didn't find out about that until after it was already in print, so I can't fix it. But I like this guy.

AS: Tell me about your first interaction with an astronaut. That must have been exciting.
AW: I wish I could remember his name. He was a space shuttle astronaut. He said, "I really liked your book and stuff, but just so you know you can actually bake the CO_2 out of lithium hydroxide canisters by heating it up." I didn't know that. There is this whole part [in my book] where Watney has to bring the oxygenator with him on his long trip. If I'd known you can just bake the CO_2 out, that would have been much simpler.

AS: Are you writing another book right now?
AW: I am. I'm working on my next book; it's called *Zhek*, it's a working title, and it's more of a traditional sci-fi. So it's not rigidly scientifically accurate, it's more soft science fiction, like aliens and stuff like that going on.

AS: I can't wait to read it.
AW: Well, thank you. ⊘

Hep Svadja

ASTRONAUT SURVIVAL TIPS
WRITTEN BY LEROY CHIAO

NASA

There is nobody better trained than an astronaut! Here are my five tips for survival.

1. MEMORIZE "BOLD FACE" PROCEDURES AND BE READY TO EXECUTE THEM
There are few things that will kill you quickly, but they are there. You do not have time to look at a checklist in these situations.

2. IF YOU ARE NOT IN A "BOLD FACE" SITUATION, THINK BEFORE ACTING. YOU CAN ALWAYS MAKE IT WORSE
If you have time, think through the consequences of your actions and how they might cause problems in other systems or areas.

3. "FLY THE AIRPLANE" FIRST
Way too often, pilots fly into the ground while troubleshooting a relatively minor problem. In any emergency, remember to keep control of your overall situation first!

4. TRAIN LIKE YOU FLY, FLY LIKE YOU TRAIN
Nothing beats training. Work through your plan and contingencies. Practice over and over until you have it cold. This way, you should be able to do your job, even if you are dizzy after orbit insertion or after entry and landing.

5. KEEP THE BIG PICTURE
Remember the overall objective! Don't get stuck in the weeds and lose perspective. This is key to mission success.

LEROY CHIAO served as a NASA astronaut from 1990-2005. During his 15-year career, he flew four missions into space, including flights aboard space shuttles Columbia, Endeavour, and Discovery. On his fourth mission, Dr. Chiao flew as co-pilot of the Russian spacecraft Soyuz TMA-5, and served as the Commander and NASA Science Officer of ISS Expedition-10.

YOUR OWN MARS MISSION

WRITTEN BY
KEITH HAMMOND

Going to Mars is humanity's straight-up toughest engineering challenge yet.
Sharpen your skills with these DIY projects, and maybe you'll get picked for the colony.

1. DIY ION THRUSTER

Build a real ion engine that flings out charged atoms for thrust, just like NASA does. Danger: This is a high-voltage project. You'll need to work safely with a neon sign transformer (NST) and they're no joke. But ion drives are probably the only way we're accelerating to Mars anytime soon —they eject ions 10 times faster than chemical rocket exhaust. Pick up some copper pipe and follow Alexander Reifsnyder's how-to at makezine.com/go/ionic-thruster.

Or start smaller (and safer) with Simon Quellen Field's tiny ion motor, using a soda-can Van de Graaff generator or an old CRT television screen as your high voltage source, at makezine.com/go/ionmotor.

2. MARTIAN SOIL GARDENING

When your life depends on growing spuds in space, you'll want to practice here at home. Buy a $500 bucket of Martian regolith simulant (as you humans call it) and develop your green thumb for the Red Planet. Formulations known as JSC Mars-1 (from a Hawaiian cinder cone) and MMS (from the Mojave desert) mimic the chemical, magnetic, and mineral properties of Martian dirt: about half SiO_2, with loads of iron, aluminum, magnesium, and calcium oxides. In experiments, tomatoes and wheat liked it fine but legumes bit the dust. What will you grow? orbitec.com/store/simulant.html

3. SATELLITE COMMUNICATIONS

If you can't talk to orbiting comsats, your pitiful calls for help will never reach Earth. Build your own Yagi radio antenna with Diana Eng for listening to satellites (makezine.com/go/yagi-antenna). Then bone up on your satcoms with Mikal Hart's Skill Builders on Iridium (makezine.com/go/iridium) and GPS (makezine.com/go/gps). (And check out the amazing tale of "Rebooting a Forgotten Satellite" on page 46 of this issue.)

4. SOLAR PANELS

What good's a radio without power? When it's time to MacGyver that SOS uplink, you'll need to squeeze every last electron out of the weak Martian sunshine. Build your own 20-watt photovoltaic panels and daisy-chain them to battery banks with Parker Jardine's Skill Builders at makezine.com/go/solar-panel and makezine.com/go/solar-power.

5. ROVER BOTS

A robot that can find its own way back to base and carry your unconscious carcass into the airlock is worth its weight in Martian regolith simulant. Learn to build autonomous rovers with Jason Short's self-balancing Arduroller from *Make:* Volume 45 (makezine.com/go/arduroller), using the ArduPilot UAV controller board. (Just remember Mars is a looong way from those GPS satellites.) Or hack up Sean Ragan's R/C video telepresence Mini Rover (makezine.com/go/mini-rover) and don't leave the airlock in the first place.

6. BREATHABLE ATMOSPHERE

Even a successful colony will be stuck inside icy life-support domes pressurized with artificial air, pining for a stroll someday in a balmy terraformed atmosphere. Practice making breathable air by using electricity to split water into hydrogen and sweet, sweet O_2. You can use the electrolysis rig from Tom Zimmerman's Hydrogen-Oxygen Bottle Rocket (makezine.com/go/HHO-rocket) — and use the rocket to send up one last message if all else fails. ✺

KEITH HAMMOND
is a projects editor at *Make:* and a space nerd since Apollo.

1. Alexander Reifsnyder, 2. Wikimedia Commons, 3. Diana Eng, 4. Jennifer C. Rowe, 5. Hep Svadja (left) and Sean Ragan (right), 6. Timmy Kucynda

BE A NASA SPACE EXPLORER

WRITTEN BY MATTHEW F. REYES

The Centennial Challenges program has awarded more than $6,000,000 in prize money across 20 competitions since its launch in 2005

NASA retains a power unique in federal government: a capacity to enter into "contracts, leases, cooperative agreements, *or other transactions* as may be necessary [...] with any person, firm, association, corporation, or educational institution." The agency's overseers have used this to launch special entrepreneurial programs, small business contracts, cash incentive prizes, and volunteer opportunities for the community.

The **Centennial Challenges** program is one of NASA's shining stars, and frequently has incubated the cash-winning contestants into government-contracted entrepreneurs such as the winners of the **Astronaut Glove**, **Lunar Lander**, and **Power Beaming** challenges.

NASA announced two new challenges in 2015:

■ **3D Printed Habitat Challenge:** Revealed at Maker Faire Bay Area, this competition offers a $2.25 million total purse for the design and construction of printable habitations for deep space exploration.

■ **CubeQuest Challenge:** This competition awards $5 million to teams that can design and build flight hardware capable of various operations around and beyond the moon. One contest, called "Last Cubesat Standing," will be won by the maker of a cubesat that can communicate back to Earth from the farthest distance in deep space.

Some challenges have been proven especially difficult. The **Sample Return Challenge** is a competition to develop a fully autonomous robot capable of sensing its environment and avoiding obstacles without compass or GPS navigation, and identify and collect rock samples to be returned to its starting point. It is in its fourth year of competitive trials and no prize money has yet to be awarded.

NASA also offers smaller-scale programs as well.

The **NASA Tournament Lab** is an actively updated portal with opportunities to solve software code-based problems.

The **International Space Apps Challenge** is a very popular, irregularly scheduled hackathon where remote teams solve a range of exploration-themed projects. One notable winner converted NASA proprietary formatted "VICAR" image files from deep space probes into more common formats — entire databases of previously unshared pictures were instantly available for public consumption!

And the agency has many more entrepreneurial-based programs, including **Small Business Innovative Research Grants** — used recently by the private group that put a 3D printer on the International Space Station. Citizen space Makers can find these programs on the NASA website: nasa.gov/solve. ✪

A tetherman prepares to catch LaserMotive's climbing robot at the 2009 Power Beaming Challenge. They were awarded $900,000 for first place.

First-place (and $200,000) winner Peter Homer demonstrates his glove at the 2009 Astronaut Glove Challenge.

MATTHEW F. REYES likes nothing more than sharing his exploration of Earth and space with NASA and helping others do it too. Find him at GoPro, Burning Man, Campus Party, and Maker Faire.

Masten Space System's lander, Xoie, descends over Pad B at the 2009 Lunar Lander Challenge in Mojave, California, winning $1,000,000.

DIY SPACE PROGRAM

Makers are building their way to outer space right now — and so can you

REBOOTING A FORGOTTEN SATELLITE

In 2014 a small cadre of international hackers set out to reconnect with a 36-year-old spacecraft that had been abandoned by NASA. Originally launched in 1978, ISEE-3 was built with simple sequence relay integrated circuits to observe the effect of solar activity on Earth's magnetic field.

After crowdfunding $160,000, the team found the probe's old operating manuals and retired experts, and developed modern software-defined radio components as they re-learned "Disco-era" satellite command and control protocols. And they built their mission control center in an abandoned McDonald's in Mountain View, California.

The collaborators managed to listen, command, and control the spacecraft for several weeks — an achievement that impressed even the old dogs at NASA. Then, on July 24, 2014, the aging thruster systems ran out of gas, preventing the team from changing the probe's course. On September 25, the spacecraft ceased operations, likely due to its reliance on solar power and ever increasing distance from the sun.

Not all hopes were dashed by the antiquated systems, though. At that year's World Maker Faire, ISEE-3 Project Co-Lead Keith Cowing highlighted the important realization that even after 40 years in deep space, the solar cells and integrated circuits were working almost as well as when they were first launched. This indicates that long-duration exploration of the solar system is possible today with inexpensive, easily accessible electronics. –*Matthew F. Reyes*

THE REAL MARTIAN FARMERS

Growing crops in Martian-spec regolith (page 44), "Martian Soil Gardening") is just one aspect of space-farming research. Plant scientists Anna-Lisa Paul and Rob Ferl, professors at the University of Florida, have been recreating alien conditions to study how plants adapt and live in space.

As a research assistant in their laboratory, I learned how to manipulate plants in microgravity flying parabolas aboard NASA's KC-135 "vomit comet," and grew plants in lower pressures and in different gas compositions, including those found on Mars.

One interesting discovery happened by accident at the University of Guelph's unique light-and-pressure lab. After a very late night setting up a gas composition experiment, my colleague Jordan Callaham and I discovered that the temperature was right, but the pressure was set too low: to nearly Martian pressure! We threw out the brown, dehydrated plants and restarted the experiment.

The next morning, we were surprised to find the affected plants had green in their leaves and life in the cells. We realized: If a greenhouse on Mars were to break during a dust storm or other incident, quick actions by an enterprising Maker Astronaut could salvage the crops and save the crew. –*MR*

COPENHAGEN SUBORBITALS

Cameron Smith's career as a prehistorian may have him studying the past, but that hasn't stopped him from looking to the future. In 2008, he decided he wanted to find a way "to fly as high as [he] could from the surface of the Earth with things that [he] could build [himself]." This desire transformed into the goal "to build a functional pressure suit."

Despite his lack of experience in engineering, he pushed forward with this project and "by applying basic principles of research, design, and a head-banging stubbornness to solve all the basic technical issues, [he] got the suit to work." He, and the Copenhagen Suborbitals, designed a suit that was "holding pressure, regulating temperature, providing a good flow of breathing gas."

His perseverance has resulted in the creation of five pressure suits, the most recent one being the Zaphod I. He is hoping to test it out on a high altitude flight using an experimental balloon later this year. –*Nicole Smith*

DIY PROJECT:
STUNNING NIGHT SKY TIME-LAPSES

Taking pictures of the night sky is downright tricky, and it's even more difficult to get a time-lapse of the stars' movement across the sky. Learn the ideal settings involved with getting the perfect nighttime photos, from aperture to shutter speed to ISO settings, and how to assemble those pictures into a beautiful time-lapse that shows how radiant the night sky really is. makezine.com/go/night-sky-time-lapse –*NS*

DIY PROJECT:
LED STARRY SKY

Enjoy the peacefulness and brilliance of a starry night sky right in your own home with this LED setup. Your display can cover just a section of your ceiling, or you can add multiple panels to put on a stellar show. Play around with shapes and colors to construct a completely unique galaxy that you can enjoy day and night. makezine.com/go/led-starry-sky –*NS*

EVEN MORE SPACE MAKERS

These groups are working on cutting-edge projects to reduce the price of exploration, while opening up accessibility to space for everyone. –*MR*

Made in Space / Future Engineers: Created the first 3D printer for use on the International Space Station. futureengineers.org, madeinspace.us

Tethers Unlimited: Developing a NASA-funded, microwave-sized "Positrusion Recycler" for recycling of 3D printed parts aboard the ISS: makezine.com/go/positrusion

Boston University Rocket Propulsion Group: Promoting their crowdfunded hybrid rocket, capable of lifting 100lbs of payload to over 435K feet, for suborbital use: burocket.org

Ultrascope: Robotic telescope that can be built and used worldwide at a price drastically lower than commercial options. For at-home use; can be operated autonomously for NASA research. Part of the Open Space Agency. openspaceagency.com/ultrascope

KickSat II Cornell Ph.D. Zac Manchester created the smallest satellite platform, and used it for the largest-ever single deployment of satellites. A radiation glitch hindered the first attempt; he's now approved for the second launch. makezine.com/go/kicksats-second-attempt/

SpaceVR.co Developing a 3D printable virtual reality camera rig to bring the overview effect of being in space to anyone with Oculus Rift, Cardboard, or any other VR device. spacevr.co

NASA, NASA, Jev Olseh, Ron Risman - TimelapseWorkshops.com, igisha, and Boston University Rocket Propulsion Group

HOW TO DETECT
KILLER
ASTEROIDS

WRITTEN BY KEITH HAMMOND

Help save the Earth from annihilation by interplanetary projectiles

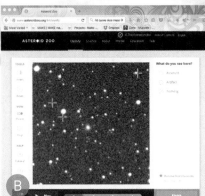

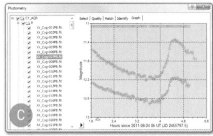

IT'S INEVITABLE — LIFE ON EARTH WILL BE INCINERATED BY A CATASTROPHIC ASTEROID STRIKE. UNLESS WE SEE THE ASTEROIDS COMING FIRST.

Scientists discover about 1,000 near-Earth objects (NEOs) every year, but that isn't fast enough — there are millions out there. In 2019, an asteroid-seeking, orbiting telescope called Sentinel (sentinelmission.org) should increase discovery 100-fold, but until then, we're flying nearly blind.

I spent a fascinating summer night at the 36" reflector telescope "Nellie" at Chabot Space and Science Center in Oakland, California, with Gerald McKeegan of the East Bay Astronomical Society as my guide. He showed me that asteroid-spotting is no cake walk, but it's totally doable. Here are four ways you can help right now.

DISCOVERY BY TELESCOPE

Point a really nice telescope overhead, and set your mount control to track the star field. With a 10" aperture, you can detect dim objects of magnitude 19, or even 20–21.

Take 5 or 10 photos with your digital camera, wait a few minutes, then take another set. Even if there's an asteroid there, you won't see it; most NEOs are just too dark. To provide enough magnitude, you've got to "stack" 5 or more photos in image processing software. Astrometrica (astrometrica.at) is designed for asteroid spotting (Figure A); it uses the FITS image format, so first convert your photos using one of the programs found at fits.gsfc.nasa.gov/fits_viewer.html.

Now look for a tiny dot that moves! If you find one, Astrometrica will automatically email a report to NASA's Minor Planet

Center (MPC). Maybe you've found a known asteroid, or space junk — or a new NEO.

DISCOVERY ONLINE

Don't have a great telescope? Surf to Asteroid Zoo (asteroidzoo.org), a joint effort of NASA and the asteroid mining company Planetary Resources. On this website (Figure B), anyone can flip through thousands of existing sky survey photos, and look for, yep, a dot that moves. Spot one and you might just be the discoverer of an asteroid previously unknown to science!

ORBIT CONFIRMATION

Scientists need more data to pinpoint orbits and speeds of known asteroids. Point your scope at an NEO chosen from the target list at minorplanetcenter.net/iau/mpc.html. Then analyze your photos in Astrometrica and report back to the MPC. McKeegan and I tracked asteroid K14M05P (aka 2014 MP5), discovered in June 2014 — and confirmed it's in a hazardous orbit near Earth.

ASTEROID CHARACTERIZATION

To nuke a deadly space rock, you'll need to know how big it is and how it's spinning. By analyzing an asteroid's "light curve" — fluctuations in brightness — you can figure its rotation period and even estimate its size, mass, and malformed-potato shape. Follow the guide at minorplanet. info/ObsGuides/Misc/photometryguide. htm, using Astrometrica or more powerful software like MaxIm-DL (Figure C) or MPO Canopus. Then submit your light curves to the MPC at minorplanet.info/call.html and you might just save your favorite planet. ⊘

KEITH HAMMOND is a projects editor at *Make:* and a space nerd since Apollo.

Earth Science and Remote Sensing Unit, NASA Johnson Space Center and Juliana Brown

Skill Builder

Ready to supercharge your skill set? These tried-and-true techniques and expert tips will help you conquer your next project with ease.

TIDBITS & TIPS ON ALMOST ANYTHING

WRITTEN BY JORDAN BUNKER
PHOTOGRAPHED BY HEP SVADJA
ILLUSTRATED BY JAMES BURKE

JORDAN BUNKER
is a technical editor for *Make:*. He is a polymathic jack-of-all-trades who enjoys manipulating ideas, atoms, and bits. Find him in his basement workshop in Oakland.

LINEAR VOLTAGE REGULATORS

Voltage regulators are an essential part of many projects that require a stable input voltage. Their job is to take an **unregulated input** voltage and output a **regulated voltage**, with the only catch being that the input voltage must be higher than the output voltage. If you've got a project in the works that needs a specific voltage, here are several options you may consider:

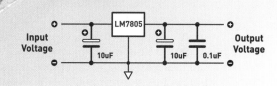

Fixed Voltage – LM78XX

The LM78XX series of linear voltage regulator chips are extremely popular, and for good reason. They're cheap, easy to use, require few other components, and have built-in circuit protection against drawing too much current. There are different models for outputting different voltages, and the last two numbers in the model number denote their voltage output. For example, the LM7805 outputs 5 volts, the LM7810 outputs 10 volts, and the LM7824 outputs 24 volts.

$$R1= V_{IN}/(\text{Max Zener Power Rating/Zener Voltage})$$

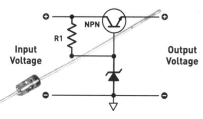

$$V_{OUT} = 1.25(1+ R2/R1) + I_{ADJ}(R2)$$

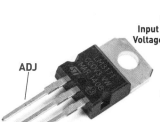

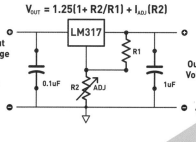

Zener Diode

You're halfway through your project, and you just realized you're fresh out of linear regulator ICs. What can you do? If you've got the right voltage zener diode and a power transistor, you can make your own fixed voltage regulator using the circuit diagram above. The output voltage will be 0.6 volts below the diode's zener voltage, due to the base-emitter voltage drop across the transistor.

Variable Voltage – LM317

When you need to be able to adjust the voltage output of a voltage regulator, the LM317 is right for you. It is very similar to the LM78XX series, except that it has an adjustment pin to change the voltage output. By adding a potentiometer to your circuit, you can use it for purposes like controlling fan speeds or variable voltage power supplies.

HEAT SINK

The larger the voltage drop across the voltage regulator, the more heat will be dissipated through the component. In order to keep it from burning up, make sure to use a heat sink!

LASERS

The word "laser" is an acronym for Light Amplification by Stimulated Emission of Radiation. Lasers are used for a wide variety of purposes, including storing data on discs, cutting through materials, and even just pointing at things. Here are two common types of lasers and how they work.

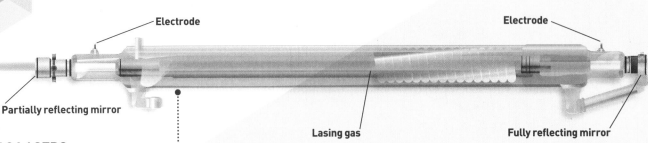

Electrode

Electrode

Partially reflecting mirror

Lasing gas

Fully reflecting mirror

GAS LASERS

Gas lasers consist of a tube of gas with mirrors on either end, one fully reflective, and the other partially reflective. When the tube is excited by an electric field, the electrons in the atoms of the gas **1** jump to a higher energy level **2**, then immediately fall back down to their original level. When they fall, the excess energy is given off as photons, producing light **3**. This light bounces between the two mirrors, which act as a resonant cavity for the light. Each pulse increases the intensity of the light, and when the light is strong enough, it shines through the less reflective mirror.

1 **2** **3**

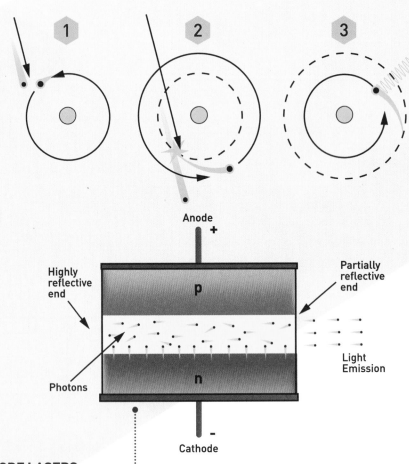

Anode
+

Highly reflective end

p

Partially reflective end

Photons

n

Light Emission

-
Cathode

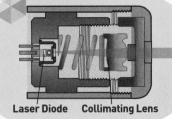

Laser Diode   **Collimating Lens**

DIODE LASERS

Diode lasers actually work much like gas lasers, except that special semiconductor materials take the place of the gas. As a result of current flowing through the semiconductors, light is generated, and resonates in a mirrored cavity until it is intense enough to shine through. The resulting beam diverges rapidly after leaving the chip, so a lens is used to collimate the light, making the light rays parallel. Using semiconductors allows lasers to be smaller and less fragile.

COLORS

Depending on the lasing medium (the gas in the laser tube, or the semiconductor materials used), lasers will produce different colors of light. This is because the electrons in different lasing mediums emit photons of different wavelengths. In the case of diode laser pointers, a laser shines through multiple materials to change the wavelength of the emitted light, resulting in different colors. The color of the laser is actually unrelated to the laser's power, and different color lasers can all have equivalent power ratings.

LASER LIGHT

Because of the way laser light is created, it is monochromatic (one color), containing only one wavelength of light. Laser light is also coherent, meaning all of the photons are in phase with one another. These properties set laser light apart from natural light, which is typically multichromatic and not coherent.

SAFETY

Not all lasers are created equal. Some lasers are safe for use by anyone, while others can cause severe burns or blindness if misused. Two major organizations have created classifications for lasers based on their potential to cause harm: the Food and Drug Administration (FDA), and the International Electrotechnical Commission (IEC).

CLASS		HAZARD	EXAMPLE
IEC	FDA		
Class 1, 1M	I	**"No hazard during normal use"** Safe for viewing in normal conditions, but may be hazardous to view through optical aids, such as binoculars or telescopes.	laser printers, CD players, DVD players
Class 2, 2M	II, IIa	**"Do not stare into beam"** Visible lasers which are safe for accidental* viewing in all conditions, but may be unsafe to stare into the beam, or to view through optical aids.	bar code scanners
Class 3R	IIIa	**"Avoid direct eye exposure"** Lasers in this class are considered low risk, but can be hazardous if directly viewed, or when staring directly into the beam. The risk of injury increases if the laser is viewed through optical aids.	laser pointers
Class 3B	IIIb	**"Avoid all direct exposure to beam"** Lasers in this class can be a hazard to eyes and skin, but viewing of the diffuse reflection is safe.	laser light show projectors, research lasers
Class 4	IV	**"Avoid eye or skin exposure to direct or scattered light from beam"** The highest classification, this light is capable of setting fire to materials, and even viewing the diffuse reflection of the beam may be dangerous.	industrial lasers, surgery lasers

* In case of accidental exposure, a person's blink reflex will limit exposure time to around 0.25 seconds.

CAUTION

LASER RADIATION
DO NOT STARE INTO BEAM

CLASS 2 LASER

All class 2 and higher lasers require a label similar to the one shown above.

ACRYLIC

Acrylic is a wonderful plastic that can be used for all sorts of different projects. It comes in both transparent and colored options, and can be machined, laser cut, or heated and bent into almost any shape.

TYPES OF ACRYLIC

Acrylic comes in two varieties: **extruded** and **cast**. While they may look identical, there are reasons you may choose one over the other based on your fabrication plans.

EXTRUDED ACRYLIC SHEET

Extruded acrylic has a lower melting temperature than cast acrylic, which makes it ideal for vector cutting on a laser cutter. This same property, however, makes it less ideal than cast acrylic if you are milling or drilling it.

CAST ACRYLIC SHEET

When laser engraving, cast acrylic is preferred, as the resulting engraving will be a frosty white color that contrasts against the rest of the acrylic. Laser engraving extruded acrylic will result in a clear engraving that doesn't contrast as well.

BENDING ACRYLIC

Using a strip heater, acrylic can be heated and bent at different angles. Store-bought strip heaters can be quite expensive, but there are ways to make do without them. If you have a toaster oven, set it to about 200° and leave the door slightly ajar. Place the part of the acrylic to be bent just above the open door, wait for it to soften enough, and then bend it to the angle desired. If you don't have a toaster oven, you can (carefully!) use a heat gun to heat the acrylic. When using these makeshift methods, bend the acrylic against the edge of a piece of wood or metal for straight, clean corners.

FLAME POLISHING

If the edge of a cut piece of clear acrylic is frosty, and you'd like to make it clearer, it can be flame polished. While it takes some practice to get right, slowly brushing the flame from a propane or MAP gas torch across the edges to melt them slightly can give them a transparent, polished appearance. Just be sure not to polish edges that must be glued, as the resulting joint may not be as strong.

BENDING SMALL ACRYLIC PIECES

If you're bending a small, thin piece of acrylic, you can use the shaft of a soldering iron as a primitive strip-heater. Place the iron handle-first into its holder, plug it in, and then use a set of third hands to hold the strip just above the iron. Make sure you keep an eye on it, as you don't want the acrylic to droop onto the soldering iron.

CUTTING ACRYLIC

For straight cuts in acrylic, a plastic-scoring blade can be used. With a straightedge as a guide, pull the blade toward you, leaving a score mark. Score the acrylic several more times along the same line, then place the acrylic on the edge of the table and use light, quick pressure to snap the piece in two.

You can also cut acrylic with more traditional blade tools such as a jigsaw, band saw, or table saw. High tooth-count plastic-cutting blades are available for these tools, and are recommended.

GLUING ACRYLIC

Acrylic is typically glued using solvent-based glues, such as Weld-On 4. Unlike many other gluing processes, acrylic glue softens the surfaces of the acrylic and welds them together, chemically bonding the two pieces into one.

To glue acrylic with solvent glue, typically a squeeze-bottle applicator with a needle tip is used. Put the acrylic on the edge where you want it, and place the needle of the glue-filled applicator where the two pieces meet. Lightly squeeze the applicator while pulling it toward you. Capillary action will draw the glue into the joint. Hold the pieces in place for several minutes (a jig with clamps works great for this), and then allow the glue to set for 10-15 minutes before moving it. After 24-48 hours, the joint will cure to full strength.

Using solvent-based glues on laser-cut extruded acrylic can cause cracking due to the internal stresses from temperature differences in the acrylic. To guard against cracking, place the acrylic on a flat sheet of glass in a shop oven (not the one in your kitchen!) at about 180°F (82°C) for about 1 hour per mm thickness, then let them air cool. This will anneal the acrylic and relieve the built up stress.

DRILLING ACRYLIC

If you try to drill a sheet of acrylic with conventional metal or wood drill bits, there's a chance you'll end up cracking it. If the sheet is thin, use a step drill bit instead. If you're working with a thick piece of acrylic, you can use conventional drill bits if you first place a piece of masking tape over the area to be drilled. If the hole is especially thick, spray some WD-40 to act as a lubricant. This will help remove chips and dissipate heat as the hole is drilled.

PROTOBOARD CIRCUITS

If you're ready to finalize a breadboarded circuit, but not quite ready to create a printed circuit board, then your next step is protoboard. There are multiple varieties of prototyping board, but the two most common are referred to as **perf board** and **stripboard**. Both are made from a flat sheet of resin with a grid of holes drilled in them, however the conductive copper on the underside is different.

PERF BOARD
When most people talk about perf board, they are referring to the ever-popular "pad-per-hole" type of protoboard. As you might suspect, each hole is surrounded by a copper pad. Connections are typically made by either bridging solder from pad to pad, or by running wires (or the bent leads of the components) from pad to pad.

Although these joints may look lumpy, it is actually quite normal for bridged solder pads to look this way.

STANDOFFS
Standoffs are often overlooked, but they are a very useful way to mount your protoboard inside a case and make your project look a bit more professional. Standoffs are also vital when you are mounting your board in a metal enclosure, as you don't want the component pins to short out on the enclosure.

PLANNING

If you've never created a complex circuit on protoboard, it can seem a little daunting. Thankfully, there are a couple of ways to lay out your design before you get started, which will make the build a lot easier.

The low-tech way to plan out protoboard circuits is with graph paper. Draw the outlines of your components where you'd like to place them on the protoboard, with the pins drawn on the intersections of the lines. Each line intersection is considered a hole. After "placing" your components, draw your connections along the lines between the components. For a tidy design, try to draw your lines straight or at 45° or 90° angles. If you need to use a jumper wire, draw it in with a different color so that it stands out. If you're using stripboard, draw circles to denote where you will break the strip.

A more modern way to plan out protoboards is with software. There are a number of different board-planning applications, but we highly recommend Fritzing, which allows you to lay out circuits on both perf board and stripboard. It's also free and runs on Windows, Mac OS X, and Linux! To download a copy, go to fritzing.org.

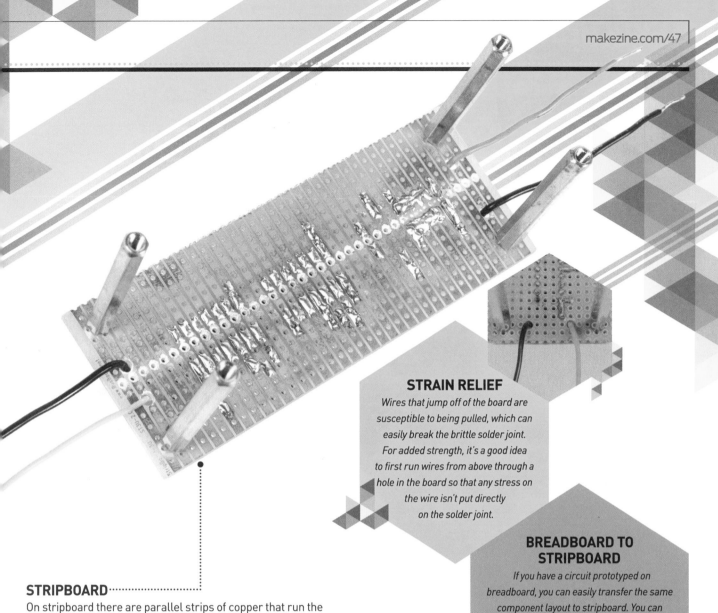

STRAIN RELIEF

Wires that jump off of the board are susceptible to being pulled, which can easily break the brittle solder joint. For added strength, it's a good idea to first run wires from above through a hole in the board so that any stress on the wire isn't put directly on the solder joint.

BREADBOARD TO STRIPBOARD

If you have a circuit prototyped on breadboard, you can easily transfer the same component layout to stripboard. You can buy stripboard with strips of copper in the same orientation as breadboard, or you can break the long copper strips down the middle into two strips. It's a quick way to make a temporary circuit more permanent.

STRIPBOARD

On stripboard there are parallel strips of copper that run the length of the board, connecting the holes together. These strips serve to connect multiple components together without wire, and can be separated, as pictured, into smaller segments with a few twists of a drill bit (see "Breaking Stripboard Traces," below right). With some clever component positioning and strategic cutting, this can eliminate a lot of wiring on the underside of the board, and works great for simple circuits.

CUTTING PROTOBOARD

Most protoboard is made out of paper laminated with phenolic resin. This material is great at resisting the heat required for soldering, but it can be difficult to cut without cracking it. Boards can be cut with traditional blade tools like a band saw or scroll saw, but it is often easier to use the score-and-snap method. Using a straightedge as a guide, score both sides of the board with a sharp knife, then place the board on the edge of the table and snap it along the marks. If you make the score marks through the center of a line of holes, there will be less material to snap, and you'll have better luck avoiding cracking.

BREAKING STRIPBOARD TRACES

An important part of using stripboard is strategically dividing up the strips of copper on the bottom to separate connections. While there are fancy purpose-made tools to break these connections, you can easily use a 4mm (or 5/16") drill bit just as well. Place the drill bit tip into the hole where you want to break the connection, and twist until the copper is cut away completely. You have just turned one strip into two separate ones!

LUNCHBOX
LASER SHOW

Break out that soldering iron, because it's time to build a laser show in a lunchbox! The Laser Show Lunchboxen project from *Make:* Volume 20 is the perfect build to practice using the tools and materials covered in this section. With three different laser light effects to choose from, your next party is sure to be a good one.
makezine.com/go/lunchbox-laser-shows

ACRYLIC   PERF BOARD

LASERS   VOLTAGE REGULATOR

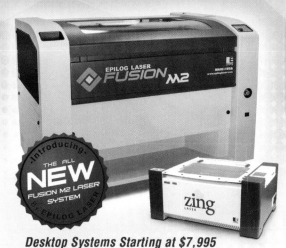

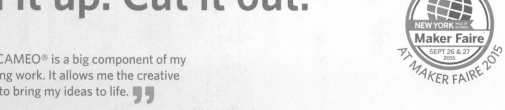

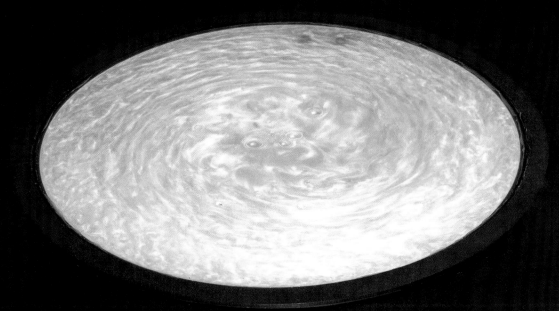

Rheoscopic Disc
Written by Ben Krasnow
Coffee Table

Build a
mesmerizing,
fluid-filled
turntable that
shows turbulent
flow patterns

BEN KRASNOW works at Google[x]
creating advanced prototypes. He
previously developed virtual reality
hardware at Valve. Follow Ben's personal
projects on his YouTube channel,
Applied Science: youtube.com/bkraz333

WHEN I LAST VISITED THE CARNEGIE SCIENCE CENTER IN PITTSBURGH, I SAW A REALLY BEAUTIFUL INTERACTIVE FLUID EXHIBIT demonstrating the swirly, churning patterns of turbulence. The fluid was contained in a 3-foot-diameter disc, about 2 inches deep with a sturdy glass top. A large Lazy-Susan bearing supported the disc so that visitors could spin it with their hands, creating a vortex in the fluid within. The effect was mesmerizing. People would spend several minutes intently watching the shimmering pearlescent fluid as they rotated the disc.

When I got home, I immediately began building a similar disc that would fit into my custom wooden coffee table. I showcased this table at Maker Faire Bay Area, and in this article I'll describe how you can build your own fluid turbulence disc from readily available parts in just a few hours spread over a weekend (Figure Ⓐ).

VORTEX VISUALIZERS

The fluid within these discs is a mixture of water, colorant, and finely ground crystals. The crystals are microscopic, flat, shiny flakes that orient themselves along the flow direction of the fluid, like logs flowing down a river. In a turbulent fluid — think of river rapids — the flow direction is chaotic, with eddy currents and swirls causing the fluid to rapidly change direction without apparent order. The flat crystals track all this chaos so you can see it — in some regions of the fluid, they're oriented so that they collectively reflect a lot of light, appearing brighter, and in other regions, they face away and reflect less light. The fluid shimmers as it churns, and is called *rheoscopic*, meaning "current showing."

A vigorous spin of the disc will stir up the fluid and the crystals will remain floating for 5 minutes or so. The vortex patterns that are formed are reminiscent of storm clouds swirling in the atmospheres of Saturn and Jupiter. They also remind me of spiral galaxies with their twisted starry arms, though those currents are primarily shaped by gravity.

Gravity does not play a big role in the patterns formed in the fluid disc. Instead, it's the friction between the liquid molecules, which are in intimate contact with each other and the walls of the container. When a fast-moving region of fluid encounters a slower-moving region, the molecules slide past each other, and their paths meander like a person trying to make their way through a crowded room.

What's interesting is that the behavior of these sliding molecules can take one of two very distinct paths — calm, laminar flow, or turbulent flow — which can be visualized in your fluid disc. Let the disc settle, then start spinning it at a constant rate. You'll notice that the fluid near the center remains calm, and the outer area near the edge is turbulent. However, there's not a gradual transition from calm to turbulent — rather there's a distinct transition point. This occurs because once a fluid starts to become turbulent due to very minor, random perturbations, it causes nearby fluid to also become turbulent, which spreads even more turbulence. The transition point forms where the smooth, orderly flow cannot resist the positive-feedback loop of perturbations that are causing chaos.

Engineers have learned how to predict when a fluid will transition from laminar flow to turbulent flow based on the fluid's viscosity, speed, and the size of the container or object that's touching the fluid. This allows the performance of many everyday objects to be optimized. For example, the dimples on a golf ball allow it to travel with less air resistance because the dimples create turbulence, which paradoxically affects the ball's flight less than laminar airflow would. If you're designing a ship, an oil pipeline, an airplane or helicopter, or many other objects that encounter fluid flow, understanding the transition to turbulence is important.

The fluid disc you'll make in this project is very similar in principle to the tools that researchers use to study fluid dynamics in the lab. It also happens to be visually entrancing — a great demonstration of the beauty of science either as a standalone presentation or, like mine, mounted in a coffee table .

1. CUT AN ACCESS HOLE

Place the Oxo turntable upside down, and choose any of the 4 holes near the periphery. Using the hole as a starting point, cut a ¾"×1½" rectangular opening in the base, with a Dremel or similar tool, as shown (Figure Ⓑ).

2. DRILL THE FILLING HOLES

Place the turntable right side up, and drill two ⁵⁄₁₆" holes in the top, spaced 1" apart, with their edges ¹⁄₁₆" away from the peripheral wall of the turntable. Don't drill through the base! It's also important to drill these holes as cleanly as possible,

Time Required:
A Weekend
Cost:
$70–$85 plus table

Materials
» **Oxo Good Grips Turntable, 16"** Amazon # B000QA2IJ0
» **Acrylic disc, 16"** TAP Plastics #01969, tapplastics.com
» **E6000 adhesive, clear** TAP Plastics #14184
» **Pearl Swirl Rheoscopic Concentrate** Steve Spangler Science #WSRL-100, stevespanglerscience.com
» **Well nuts, #8-32, for ⁵⁄₁₆" hole (2)** McMaster-Carr #93495A130, mcmaster.com
» **Machine screws, #8-32 × ⁷⁄₁₆" (2)** McMaster #91772A193
» **Funnel, micro size** McMaster #1479T82
» **Distilled water, 1gal**
» **Liquid food coloring**
» **Spray paint, black**

Tools
» **Drill and ⁵⁄₁₆" bit**
» **High-speed rotary tool** such as a Dremel, with a bit for cutting plastic
» **Screwdrivers: Phillips and small flat-head**
» **Craft knife** with new blade

OPTIONAL, FOR MOUNTING IN TABLETOP:
» **Jigsaw**
» **Circle drawing tool**
» **Drill bit, ³⁄₈" or bigger**
» **Masking tape**

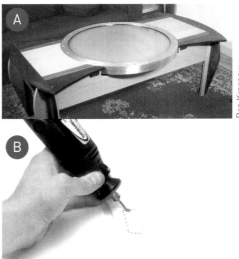

Ben Krasnow

IMPORTANT: Be sure to only cut through the base of the turntable, not the turntable top itself.

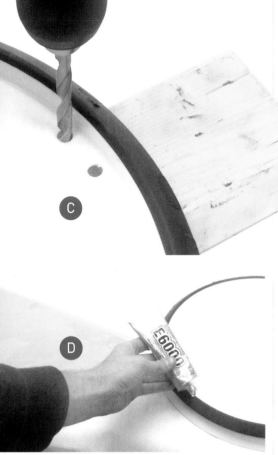

C

D

IMPORTANT: Do not substitute another adhesive without first testing its ability to make a very firm bond between the acrylic and the rubber turntable edge. A failed adhesive bond may cause a sudden, large leak.

E

so keep the drill bit perpendicular and make sure it doesn't wander (Figure **C**).

Use the craft knife to neatly cut away any burrs. Clean the turntable top of all dust and label residue.

3. ATTACH THE ACRYLIC DISC

Apply a continuous, consistent bead of E6000 adhesive to the turntable's rubber edge. The glue should completely cover the highest point on the edge. It helps to rotate the turntable while holding the glue tube in one place and squeezing slowly (Figure **D**). Apply the adhesive liberally — the mess will be hidden by later steps.

Remove one of the protective films from the 16" acrylic disc, and slowly lower the disc straight down onto the adhesive, keeping it centered on the turntable by eye. Apply very light pressure to seat the disc onto the rubber (Figure **E**).

If there are any gaps in the adhesive bond, you can fix them later — do not reposition once contact with the adhesive is made! Clamping is not needed.

4. INSPECT AND TOUCH UP ADHESIVE

After a few hours, inspect the adhesive bond for gaps or thin areas that might leak. If you find any, apply more adhesive with a pencil or chopstick (Figure **F**). Make sure the adhesive forms a continuous seal between the disc and turntable. Messy is OK. You'll cover it up later.

Let the adhesive cure at least 24 hours, or 48 if the ambient temperature is cool.

5. ADD THE LIQUID

Pour out 1 cup of the distilled water to make space within the jug. Shake the bottle of Pearl Swirl concentrate vigorously, then add half the bottle and a few drops of food

coloring to the water jug. Invert the jug several times to mix. I used 4 drops of blue coloring for my disc.

Position the turntable upside down and rotate the base so that your rectangular access hole lines up with the drilled holes. Prop up one edge of the turntable with a 1" wood block so that the drilled holes are the highest point of the sealed space. Use the small funnel to fill the disc from the jug (Figure **G**). Periodically stop and shake the jug to keep everything mixed well.

6. CLOSE IT UP

When the disc is completely flooded and water spills from the holes, insert a well nut into each hole, then place a screw into each nut, and lightly tighten. Use a flat screwdriver to keep the well nut from spinning as you get the screw started (Figure **H**). The screw doesn't need to be very tight.

Turn the disc right side up and check for air bubbles. To eliminate bubbles, put the disc back into the filling position, remove one well nut, and try slightly different angles on the wood block so that all air bubbles eventually exit.

7. CUT THE PROTECTIVE FILM MASK

With a very steady hand propped on a wood block, place the knife tip 1" in from the edge of the disc, and slowly rotate the disc with the other hand (Figure **I**). Use very light pressure with the knife, so that the protective film is cut, but the acrylic is not harmed too much.

Peel away the outer 1" strip of protective film so that only the outer edge of the acrylic is exposed (Figure **J**).

8. SPRAY-PAINT

Now you'll paint the exposed acrylic to cover up all your messy glue. Apply 2–3 light coats

Ben Krasnow

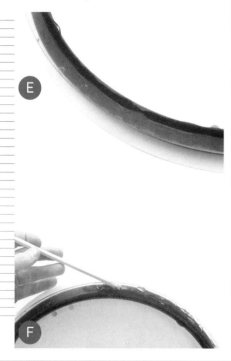

F

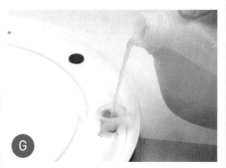

G

H

of spray paint. I chose Krylon flat black.

When it's dry, peel away the rest of the film (Figure ⓚ), and your disc is done.

9. MOUNT IT IN A TABLETOP (OPTIONAL)

For this project, we like the Kragsta table ($89) from Ikea, in black. We mounted the disc "proud," with the acrylic floating about 1mm above the tabletop for clearance. Set your circle drawing tool to a radius of 8", place the needle in the existing centered hole on the bottom of the tabletop, and draw the circle. Drill a hole ⅜" or bigger inside the circle, then insert your jigsaw blade and carefully cut out the circle.

Temporarily attach the leg supports, and mark where the circular hole meets them. Detach them and remove ⅝" depth of material between your marks, to accommodate the turntable base. Mask and spray-paint all cut edges.

Then assemble the table, set your rheoscopic disc into the tabletop, and take your new vortex visualizer for a spin! ◑

J

I

K

More Fluid-Filled Fun from MAKEZINE.COM

LAMINAR FLOW WATER FOUNTAIN
DIY high-flying fountains a la Disney or Bellagio.

GLOWING DANCING OOBLECK
Bizarre, lifelike motion from a non-Newtonian fluid.

TANTALUS CUP
Decipher the siphon and build a high-class practical joke.

Get more photos and build tips, and share your own rheoscopic disc creations at makezine.com/go/rheoscopic-disc-coffee-table.

Hep Svadja

DIY
Aquaponic Garden
with Arduino

Grow fish, fruits, and veggies together in this compact, automated system. You can even control it via the internet.

Written by Rik Kretzinger
Illustrated by Matthew Billington

LOTS OF CREATIVE AQUAPONICS SYSTEMS ARE BEING DEVELOPED FOR RAISING FISH AND VEGETABLES TOGETHER IN SMALL SPACES. UNFORTUNATELY — FOR YOU AND THE FISH — most of these systems fail to address a number of common problems and so they end up on Craigslist or in the trash. And none of them are automated effectively at an affordable price.

I set out to build something better: a smart, small-footprint DIY aquaponic system controlled by an Arduino and built with parts from your local big box store or Amazon — OK, except the valve, that's from eBay.

The Aquaponic Balcony Garden can be fully automated, with relay-controlled pumps, and sensors to detect humidity, temperature, soil moisture, and water level in the fish tank. It's got a backup air pump to save your fish if the power goes out, and a master system kill relay in case anything goes wrong. You can even operate it via the internet. I'm developing kits for sale, but I'm also sharing the complete DIY instructions and Arduino code, so that anyone can build their own.

Notably, this garden uses no bell siphons, which are prone to failure. Instead, the grow bed is watered by a motorized ball valve that allows for gravity feed pressures. This gives you precise control over water cycles so you can schedule them into a grow plan to accommodate a large range of plants.

RULES OF AQUAPONICS

Three basic rules in aquaponics are important to the health of the system:

» **1:1 Relationship** — between fish tank volume and grow bed volume.

» **Fish Stocking Density** — 1 pound (500g) of fish for every 5–10 gallons (20–40 liters) of fish tank water.

» **Feeding Fish** — Only feed the amount they can eat in 5 minutes.

Most small systems (and their owners) break all three rules, causing the system to fail or never reach a balance.

I designed the Aquaponic Balcony Garden with a buffer — the DIY bio-reactor — so it will still work even for newbies who are breaking rules. Inside the bio-reactor, small plastic pieces called "moving bed media" provide maximum surface area for the growth of beneficial bacteria that remove ammonia and nitrites from the water. Your fish will thank you.

THREE WAYS TO PLAY

You can configure the Balcony Garden three ways, depending on your experience level:

» **Basic**: A simple timer performs all functions needed for a stand-alone aquaponic garden.

» **Local microcontroller**: An Arduino Uno microcontroller provides precise control of cycle times, and collects sensor data to show you what's taking place in the growing environment.

» **IoT (Internet of Things)**: An Arduino Yún microcontroller lets you control your garden anywhere in the world. Receive text messages (say, when the grow bed is being filled), do data streaming and logging, and more.

In this article I'll show you how it all works. For complete step-by-step instructions, visit the project page online at makezine.com/go/aquaponic-balcony-garden.

WHY DO AQUAPONICS?

Growing fish and plants together is fun and relaxing. It's sustainable and chemical free — the fish fertilize your plants — and uses less water and electricity than running an aquarium and hydroponic garden separately. And aquaponic systems can be taken off grid very easily.

The *nitrogen cycle* is important to know when doing aquaponics. When fish digest food, their waste breaks down to produce ammonia (NH_3), which is toxic to the fish in high levels. Beneficial bacteria break down the ammonia into less-toxic nitrite (NO_2), then convert the nitrite into nitrate (NO_3), a plant nutrient that's nontoxic to fish.

RIK KRETZINGER grew up on a Christmas tree farm outside Lompoc, California, and earned a degree in horticulture from Cal Poly San Luis Obispo. He has worked building commercial greenhouses, growing tropical plants and roses, and as an ag specialist in the western states. Today he works in biotech and high-tech industries while building his aquaponics business.

Time Required: 1 Week
Cost: $475–$500

Materials

Watch for the kit at agponics.com, or buy the following parts separately to DIY. For complete vendor and part numbers, see the project page online.

STRUCTURE

» **27-gallon totes with snap lids (2)** Lowe's item #44066, lowes.com

» **Fiber cement backer board, ¼"×20"×30"** Lowe's #11640, aka Hardie board

» **Deck balusters, 33½", white composite (4)** Lowe's #44551

» **Deck post sleeve, 4×4, white composite** Lowe's #160835, measures 4.1"×4.1"

» **Medium-density fiberboard (MDF), primed, 0.688"×3½"×18"** Lowe's #85819

» **PVC board, nominal 1×6, 18" length** Lowe's #209621, measures ¾"×5½"

» **Primer/sealer, white, 1gal** such as Zinsser Bulls Eye

VALVE AND PUMPS

» **Motorized ball valve, 1", 110–240VAC, normally open, gravity feed** Ehcotech #M21SE-1-C2CW, valves4projects.com

» **Water pump, submersible inline** Elemental Solutions H2O, rated 370gph

» **Aquarium air pump** Tetra Whisper

» **Air bubbler stone** EcoPlus #728415

» **Battery-operated backup air pump** Silent Air #SAB11

» **Polyethylene tubing, ¼", 24" length** Lowe's #443061

» **Vinyl tubing, clear, ¼", 12" length** Lowe's #443063

» **Vinyl tubing, black, ¼", 2" length** Lowe's #44215

» **Barb fittings, ¼", 90° elbows (3)** Lowe's #160752

MICROCONTROLLER AND SENSORS

» **Arduino Uno or Arduino Yún microcontroller board** Maker Shed # MKSP99 or MKSP24, makershed.com

» **Proto Shield for Arduino** Adafruit Industries #2077 or #51

» **Humidity/temperature sensor, DHT-22**

» **Temperature sensors, waterproof, DS18B20 (2)**

» **Float switch liquid level sensor, right angle** Amico #A11062100UX0008

CONTINUED ON PAGE 64

» **Resistors, ¼W: 10kΩ (2), 470Ω (2)**
» **CAT-5/RJ-45 connectors: black (1), yellow (1), blue (1), and almond (1)** Leviton #41108-RE5, 41108-RY5, 5G108-AL5, and 41108-RA5
» **Ribbon cable, 4 wire, 22 gauge, 20' total length**
» **Connector receptacles, 4-position, 2.54mm (4)** Molex #22-01-3047, Jameco #234819
» **Cable crimp connectors, female (25)** Molex #08-55-0102, Jameco #234931
» **Waterproof cable glands for 4mm–7mm cable (6)** Amico #S14010200AM0866
» **ABS pipe, 2", 2" length**
» **ABS fittings, 2", "test plate" caps (2)** Canplas #103762

FILTRATION
» **Bio-filter moving bed media, 25mm x 12mm (200 pieces)** Kaldnes K3 Bio Media; buy them "bio-activated" (wholesalekoifarm.com) to save a lot of system cycling time. You can find small quantities on eBay.
» **Bio-filter pad, Matala Blue, half sheet** Matala #SMBE24, Amazon #B003CIAOZS
» **Air purifier carbon filter** Lowe's #586651
» **Polyester fiberfill, 12oz bag** such as Poly-Fil, from craft stores or online
» **PVC strainer, 4¼"** Lowe's #253139

PLUMBING
» **PVC pipe, 6", 15cm lengths (2)**
» **PVC fittings, 6": coupling (1), caps (2)** Lowe's #1815 and 122863
» **PVC fittings, 4": tee (1), adapters (4), and cleanout plugs (4)** Lowe's #24125, 24094 and 146812
» **PVC pipe, 1", about 36" total length** Your mileage may vary.
» **PVC slip fittings, 1": 90° elbows (4), in-line ball valves (6), and cap (1)** Lowe's #23870, 108956, and 23897
» **PVC nipples, threaded, 1", 2" long (2)** Lowe's #57071
» **PVC swivel fittings, 1": 90° elbows (2), O-ring MPT adapters (2), and PVC-Lock MPT adapter (1)** Amazon #B00B8CBCK6, Sprinkler Warehouse #332-010, and Home Depot #859580
» **PVC insert fittings: 90° elbows (3), 90° elbows to FPT (6)** PVCFittingsOnline.com #1406-010 and 1407-010
» **ABS slip fittings: 4"×3" couplings (4)** Lowe's #22809
» **Flexible bulkhead fittings: ½" (3) and 1" (2)** Uniseal #32324 and 32326
» **Corrugated tubing, 1", 10' length** Lowe's #63417

CEMENTS AND SEALANTS
» **PVC cement, clear** Lowe's #23467
» **PVC/ABS transition cement** Lowe's #68640
» **Lexel clear caulk, 5oz** stronger than silicone
» **Sugru moldable silicone rubber** Maker Shed #MKSUMC
» **Epoxy putty, 2-part** such as TAP Plastics Magic-Sculpt
» **Dura-Rubber coating, white (optional)** Rubberizeit #19, for fish tank. It's food-, fish-, and people-safe.

CONTINUED ON PAGE 66

THE AQUAPONIC BALCONY GARDEN — HOW IT WORKS

Grow bed, 27 gallons ❶ — holds soil or other growth media for plants. It's also got:
» **Moisture probe** — DIY analog probe made from 2 stainless screws
» **DS18B20 temperature probe**
» **Water inlet / overflow preventer ❷** — delivers filtered and bio-treated water to the grow bed, and prevents overflow if the bed is overfilled
» **Water drain / root clogging preventer ❸** — covers the outlet, and admits water but not roots. (This is a major problem in aquaponics — roots clog everything unless you design for the problem.) It's a perforated 1" pipe inside a slitted 2" pipe — just twist to snap off invading roots.

Fish tank, 27 gallons ❹ — can be painted white with a food-safe rubberized coating for appearance or heat reflection. It also supports the grow bed platform and houses the following elements:
» **Ultrasonic distance sensor ❺** — measures water level at all times
» **DS18B20 temperature probe** — measures water temperature
» **Float switch** — an analog backup to the ultrasonic sensor
» **Water outlet** — Add a screen to protect small fish from pump suction.
» **Water inlet** — from grow bed drain valve
» **Water pump** — A submersible pump is operated in *inline mode* to pass water to the grow bed from the bio-reactor, *not* from the fish tank — it's just mounted in here to keep the footprint small.

Fish tank cover ❻ — Keeps out unwanted items like leaves and hungry raccoons, while allowing light to pass through. Fish can sense day and night, and it's important to their health.

Grow platform ❼ — supports grow bed, routes all the wiring, and houses the all-important drain valve

Electric drain valve ❽ — delivers grow bed water back to the fish tank on command, for complete automation. It also opens in the event of power failure, delivering maximum water (and oxygen) to your fish.

Double leg support ❾ —The heart and brain of the garden, it supports the grow platform and houses the microcontroller and electrical connections.

In a see-through box, a lighted "**pilot switch**" ❾ⓐ controls **120V AC mains power**, so the entire system is easily switched off (before you put your hands in water). You can also control it with a relay.

In the **AC outlets** housing ❾ⓑ, 4 relay-switched outlets power the air and water pumps, optional heater, and battery-backup air pump. The **opto-isolated relays** are controlled by your Arduino, based on data from the sensors and probes, and they're configurable — normally open or normally closed — depending on your needs.

In the **DC converter** housing ❾ⓒ, a **transformer** steps down 120V AC to 12V DC, then **9V and 5V converters** regulate it for your microcontrollers and sensors. This 12V conversion also lets you use the garden off-grid, or connect solar backup power.

Finally, in the **sensor/microcontroller** housing ❾ⓓ, DC power is distributed with independent terminal blocks: 9V to the **Arduino**, 5V to sensors. A **humidity/temp sensor** helps you understand the environment your plants are growing in.

> **NOTE:** The biggest problem with this design is that *all sensors must be grounded back to the microcontroller*. If you forget to ground back, the electronics will not work, period. I'd love to hear your improvements!

Single leg support ❿ — supports the platform and houses the electrical control for the grow bed drain valve — either a relay or, for the Basic version, a digital timer.

DIY bio-reactor ⓫ — Provides maximum surface area for the growth of bacteria that convert toxic fish waste to nontoxic plant fertilizer, using an aquarium bubbler and moving bed media. This is a "big system" component not seen in other small systems.

DIY solids filter ⓬ — Another "big system" feature, this 3-stage filter captures solid waste for bacterial breakdown and settles out heavier material for removal.

Air pump ⓭ — Ordinary aquarium bubbler pump supplies air to the bio-reactor.

Connection plumbing ⓮ — to hook it all up. Your site and configuration will vary — just make sure it's all watertight.

SAVINGS VOUCHER

	MAKE Magazine	Total Value	You Get	Your Discount	Your Price
6 issues		**$69.93**	1 year of MAKE + 2015 Guide to 3D Printing PDF	**50%**	**$34.95**

☐ PAYMENT ENCLOSED
☐ BILL ME LATER

Download immediately! 2015 Guide to 3D Printing
Email address required for delivery

ENTER THE DRONE ZONE
27 PROJECTS
[FIRST-PERSON]
DRONE RACING
FAST, FIERCE, HIGH-FLYING FUN

ANNUAL GUIDE TO 3D PRINTING
26
Make:
3D PRINT YOUR CAR

Get it now! 2015 Guide to 3D Printing PDF

NAME _____

ADDRESS _____

CITY _____ STATE _____ ZIP _____

COUNTRY _____

EMAIL ADDRESS _____ *required for delivery of 3D Printing Guide and access to digital editions

One year of MAKE Magazine is 6 bimonthly issues. Please allow 4–6 weeks for delivery of your first issue. Additional postage required for subscriptions outside of the U.S. Access your digital edition instantly upon receipt of payment at make-digital.com.

MAKEZINE.COM/SUBSCRIBE
Enter Offer Code: B54NS3

BUSINESS REPLY MAIL
FIRST-CLASS MAIL   PERMIT NO. 865   NORTH HOLLYWOOD, CA

POSTAGE WILL BE PAID BY ADDRESSEE

Make:

PO BOX 17046
NORTH HOLLYWOOD CA 91615-9186

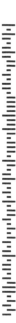

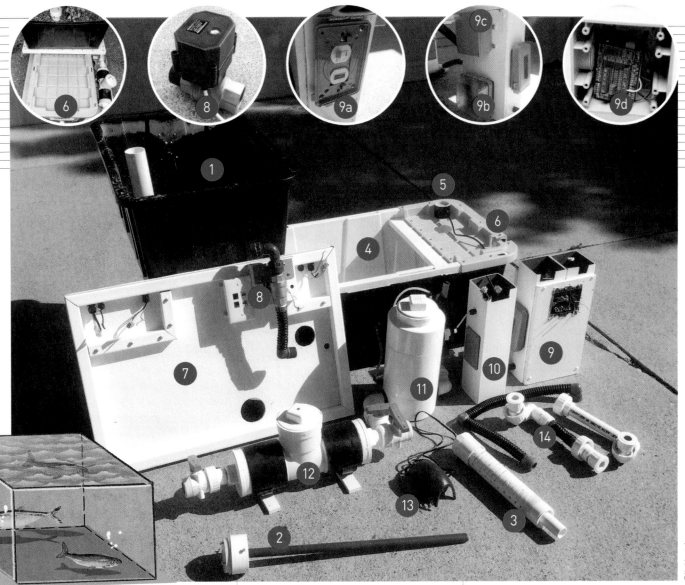

Rik Kretzinger

BUILDING YOUR AQUAPONIC BALCONY GARDEN

1. PREPARE THE GARDEN COMPONENTS

To prepare the fish tank, grow bed, platform, supports, and various plumbing and wiring bits, visit the project page at makezine.com/go/aquaponic-balcony-garden. Here I'll just explain the secret sauce — the bio-reactor and the electronics — and show how it all goes together.

2. BUILD THE BIO-REACTOR

Trace the 4" PVC adapter on top of a 6" PVC end cap, cut out the circle, and glue the adapter in. Seal the joint with Lexel.

Drill a $5/32$" hole in the 4" PVC cleanout plug, then insert a #8-32 screw. You'll loosen it to bleed air out of the bio-reactor.

Glue together the base: a 6" end cap, a

15cm length of 6" pipe, and the 6" coupler. Drill $13/4$" holes through opposite sides, $21/4$" from the bottom, and install the 1" Uniseal bulkheads. Insert inlet and outlet pipes of 1" PVC, then glue a 90° elbow and a ball valve on each. Inside, slip a perforated 1" cap onto the outlet as a coarse screen (Figure **A**).

Now glue another 15cm pipe into the top of the base, and finally your modified cap.

To make a visual water gauge, drill and tap two $27/64$" holes near the top and bottom, and mount 2 of the watertight cable glands. Connect these with $1/4$" clear tubing and elbow barb fittings. Now the tubing will show you the water level inside (Figure **B**).

Mount a cable gland on top, pass the air hose through, and put the aquarium bubbler inside on the bottom.

Finally, add 200 pieces of the K-3 moving bed media (Figure **C**), and fill with purified

HARDWARE

» **Machine screws, stainless steel, #8-32 pan head:** ⅝" (50), ¾" (20), 1" (10)
» **Lock nuts, stainless steel, nylon insert, #8** (10)

ELECTRICAL

» **Relay boards for Arduino, opto-isolated, high-current: 4-channel (1) and 1-channel (1)** YourDuino #EA-040406 and EA-040409
» **Solid-core wire, 14AWG, 600V rating: red (20 feet) and black (20 feet)** for all 110–120V wiring. It will carry 10 amps — more than we need, but much safer!
» **Hookup wire, 22 gauge** Maker Shed #MKEE3, for the low voltage stuff
» **Switching power supply, 24V DC, 2A** Amazon #B00BXXIYJ2. Transforms your AC to DC.
» **DC/DC converters: 12V to 9V (1) and 12V to 5V (1)** Amazon #B00LXTCQHM and B00CXKBJI2
» **Terminal strip, Euro style, 12-position** Adafruit #677
» **Barrier strips, 4-position dual row (2)** Amazon #B0054N4CBG
» **Terminal jumper, 8-position** Amazon #B00D0W93NU
» **Male DC power adapter, 2.1mm plug to screw terminals** Adafruit #369
» **DIY USB connector, Micro-B plug** Adafruit #1390
» **Screw terminal blocks, 2-pin, 3.5mm (12)** Adafruit #724
» **Electrical boxes with covers: 1-gang (1) and 2-gang (2)** Lowe's #130898, 37117, and 70975, with covers 330360, 79214, and 690020
» **Electrical outlets, 125V 15A, duplex (2)** Lowe's #216401 or Home Depot 697882
» **Combination switch, 15A, with pilot light** Lowe's #541764 or Home Depot #649366
» **Wire connectors, push-in: 2-port (5) and 8-port (1)** Home Depot #631091 and Amazon #B00HRHXMTU

Tools

» **PVC cutter, 1"**
» **Cordless drill, ⅜" chuck**
» **Drill bits:** ⅛", 5/32", 27/64", and ⅜"
» **Forstner bits:** 1", 1¼"
» **Hole saws:** 1½", 1¾" or 44mm, 2½", and 2¾"
» **Tape measure for sewing**
» **Combination square, 12"**
» **Thread taps:** ½"-13 NC and ¾"-14 NPT such as Vermont American #21670 and Drill America #DWTPT3/4
» **Jigsaw**
» **Wire strippers** Adafruit #527
» **Flush diagonal cutters** Adafruit #152
» **Digital multimeter, auto-ranging** Adafruit #850
» **Pliers, soft jaw, adjustable** Silverline #595757
» **Small saws, power or hand** I used my Dremel Saw-Max (rotary) and Multi-Max (oscillating) but you can use keyhole saws and your jigsaw.
» **Screwdriver, #2 Phillips**
» **Bandsaw (optional)**
» **Miter saw (optional)**

D

E

F

G

H

water (or cycled water; see Step 7). They'll float until they develop a healthy bacterial film, then they'll sink. They're self-cleaning as they bounce around in the water.

3. INSTALL THE SENSORS

Wire each sensor to a DIY CAT-5 cable and its appropriate resistor as shown on the project page. Install the humidity/temp sensor (Figure **D**) in the microcontroller housing.

Mount a cable gland in the side of the fish tank, and one in the grow bed. Route a temperature probe (Figure **E**) through each.

Drill holes to fit your ultrasonic sensor's "eyes" in the back of an ABS "test plate" pipe cap, then adhere the sensor with epoxy putty. Mount the cap in a 2" hole in the stationary fish tank lid, so the sensor can watch the water. On top, add a slice of 2" ABS pipe and mount a cable gland (Figure **F**), then waterproof it all with rubberizing compound, and seal it with a second cap.

Mount the old-school float switch in the wall of the fish tank, 4"–5" below the top. Mount the media moisture probe — really old-school, it's just 2 stainless steel screws and a resistor — in the wall of the grow bed, 3" from the top.

4. CONNECT AND TEST YOUR ARDUINO

Solder the extended headers onto the Proto Shield, then solder the terminal blocks and connect them to their respective pins as shown on the project page. Connect all sensors and relays to their terminals on the shield, then plug the shield into your Arduino. *Be sure to ground the shield* to the ground terminal in the sensor/microcontroller housing (Figure **G**).

Download the Arduino code from the project page. Then upload each sensor's test code to the Arduino, in turn, and test your sensors. Upload the relay code, and test your relays too.

5. ASSEMBLE YOUR AQUAPONIC GARDEN

Position the fish tank in a level, well-supported location with good access from all sides. Connect the solids filter to the fish tank outlet, and fill it completely with water. Place the bio-reactor behind the tank, connect it to the solids filter, and fill it too (Figure **H**).

Connect the water pump to its fitting inside the fish tank, centered beneath the pass-through hole in the lid. Connect the

Rik Kretzinger

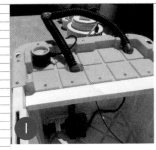

grow bed drain return hose now as well (Figure **I**).

Position the support legs on either side of the bio-reactor and set the grow platform on top of the tank. Now plug in all the electrical connectors (Figure **J**). Don't forget the drain valve connection in the single leg.

Insert each leg into its socket in the platform, then slide the platform into place against the stops on the fish tank lid. Attach the backup air pump below with velcro.

Put the grow bed on top, and connect it to the drain valve. Connect the drain return hose to the valve as well.

Finally, twist the water inlet/overflow pre-venter stack, and the root-proof drain stack, into place inside the grow bed (Figure **K**). Test the pump — you should now be successfully watering your grow bed!

You can fill your grow bed with media now. The fish tank water will get dirty; just run the pump a few hours until it clears.

6. PROGRAM YOUR GARDEN

You're ready to automate. Upload the sketch *Balcony_Timer.ino* sketch to your Arduino Uno, then plug everything in and test the entire garden. The code will run the water and air pumps continuously, and open the drain on your schedule to control the fill-flood cycle. Modify it to water more on hot days, or switch on a heater when the temperature drops.

For the IoT version, upload the sketch *Timer_SMS_Streaming.ino* to your Arduino Yún. It will stream your garden data live to Microsoft Power BI, and it will send you SMS messages via your Temboo and Twilio accounts. Learn more on the project page.

7. CYCLE YOUR GARDEN

That just means running it long enough for the beneficial bacteria colonies to become fully established. This can take 2 months with brand new water and media, or 2 weeks if you start with water and media from an already cycled system. (That's why

I'll provide start-up water with my kits.)

It's best to start out with small "feeder" goldfish and feed them very lightly. Check the ammonia and nitrite levels with a test kit purchased at your local pet store.

GET GROWING!

Your completed Balcony Aquaponic Garden (Figure **L**) can grow all kinds of fish, as food or as pets (Figure **M**), just watch your system temperatures and match fish that can live well within those values. Some good choices are:

» **Tilapia** — you'll need a heater in colder climates, and they're illegal in some jurisdictions; check with your state's agriculture or fish and wildlife department

» **Catfish** — these get big, so don't overload the tank — only a few

» **Shubunkin goldfish** — they can take just about all temperatures

» **Black moor goldfish** — another hardy breed.

Plants that will thrive in the garden (Figure **N**) include basil, cilantro, chives, peppers, strawberries, and tomatoes — just make sure you perform root maintenance.

MAINTAINING YOUR AQUAPONIC GARDEN

Once a week or so, open the cap of the grow bed drain, check for root activity, and maintain as needed (Figure **O**).

When the pump draw going into the grow bed slows down, it's time to clean the solids filter. The waste material can be deposited in other gardens or flowerpots. ◗

Get complete step-by-step instructions, code, and tips on the project page: make-zine.com/go/aquaponic-balcony-garden.

Hep Svajda

Water Rocket Launcher

Written by Mike Westerfield

Use household items to build an inexpensive water-powered rocket that can soar at 100mph

Time Required:
2 Hours from Kit*
4–6 Hours from Scratch
Cost:
$25–$45

Materials

» **Wood boards, 1×3, 12" long (2)** I used pine.
» **T-plate bracket**
» **Tube straps, ¾" (2)**
» **Hose barb, ¼" brass**
» **PVC pipe, ½" Schedule 40, 25" total length**
» **PVC slip fittings, ½" Schedule 40:** end cap (1), 90° elbow (1), coupling (1), female threaded adapter (1), and male threaded adapter (1)
» **PVC pipe, 1½", 1½" length**
» **Brass tube, ¹⁹/₃₂" OD, 0.014" thickness, 1" length** or similar
» **O-ring, ⁹/₁₆" ID, ¹³/₁₆" OD, ⅛" thick**
» **Screw eyes, large (2)**
» **Mason line, #18, 25' length** or similar strong cord
» **Vinyl tubing, ⁵/₁₆" clear, 20' length**
» **Hose clamp, 1½"**
» **Tire valve, Schrader type**
» **Hose clamps, worm gear type (2)**
» **Zip ties (20)**
» **Tent stakes (2)**

FOR THE ROCKET:
» **Plastic bottle, 2-liter soda** Other types will not withstand the pressure.
» **Foamcore board** enough for 3 fins
» **Pipe insulation, 1" length**

Tools

» Drill and bits
» Saw
» Wrench and screwdriver
» Needlenose pliers
» Utility knife
» Scissors
» Rubbing alcohol
» WD-40
» Duct tape
» **Epoxy** quick-setting is best
» **PVC cement and primer**
» **Teflon tape**
» **Polyurethane glue,** such as Titebond or Gorilla Glue
» **Rubber band**
» **Sandpaper**
» **Bike pump or compressor**

***The Water Rocket Launcher Kit** (#MKBW01) **is available in the Maker Shed. If you have the kit, skip ahead to Building a Water Rocket.**

This project and many more can be found in our book *Make: Rockets, Down to Earth Rocket Science.*

Make: Rockets
Down to Earth Rocket Science
By Mike Westerfield

Hep Svadja

A

B

MIKE WESTERFIELD
runs the Byte Works, an independent software publishing and consulting firm. He is a PADI scuba instructor who lives in Albuquerque with his wife, enjoying being an empty nester and spoiling his grandchildren.

WATER ROCKET LAUNCHERS ARE EASY TO BUILD IF YOU HAVE A FEW HOUSEHOLD TOOLS. And they are a blast to launch too; the hardest part is holding the rocket on a pad while it is under as much as 75psi of pressure. This launcher fixes that — it uses a high pressure O-ring and a zip-tie mechanism to secure the rocket until blastoff. It also has a mechanism for releasing the pressure remotely, so you never have to approach the pressurized rocket.

1. CUT THE LAUNCH TUBE
Place a ½" Schedule 40 PVC pipe into the bottle about 2" from the bottom. Mark and cut the exposed pipe about ½" from the neck of the bottle — giving you a roughly 10½" piece (Figure A).

> **NOTE:** Most 2-liter bottles fit perfectly on a ½" PVC pipe, but a few are too tight, so you will need to experiment a bit.

2. ROUND THE END
Sand the end of the PVC pipe that will slide in and out of the bottle.

3. CUT THE LAUNCH TUBE FOR THE O-RING
Dry-fit a ½" PVC adapter and insert the pipe fully into the bottle. Find a spot in the bottle's neck where it does not change diameter, then remove the launch tube and cut the pipe here, typically ½" in from the bottle opening (Figure B). These cut ends will serve as the seat for the O-ring — the critical component for holding pressure — so make sure your cuts are absolutely straight, and sand them flat if necessary.

4. CUT THE REST OF THE PVC PARTS
You'll need 3 more pieces from the remaining length of ½" schedule 40 PVC pipe, roughly 1", 4", and 9" long.

5. PREPARE THE SLIP COLLAR
Take your 1½" length of 1½" PVC pipe and drill two ³/₁₆" holes on opposite sides of one end (Figure C). You'll l tie the launch string through these holes.

Notice the orange duct tape — in Step 12 you'll apply duct tape inside the collar as needed to fit your bottle and zip-ties in the launch mechanism.

6. CREATE THE O-RING SEAT
Rough up the ends of the brass tube — but not the middle, where the O-ring will sit. Use epoxy to glue this tube inside the 2 sections of the PVC launch tube you cut in Step 3, so the gap between the PVC parts is exactly the width of the O-ring. Once the glue dries, stretch the O-ring over the tube and slide it into place (Figure D).

7. ASSEMBLE THE LAUNCH TUBE
Place the launch tube into a slip coupling as shown, followed by the 4" piece of ½" PVC and the male slip adapter (Figure E). Glue these together using PVC cement or epoxy.

8. MOUNT THE HOSE BARB
Drill a ½" hole in the center of the PVC end cap and thread a brass hose barb into it, being careful to keep it straight (Figure F). Since this pressurizes the launcher, remove the barb and inspect the cap for cracks. If it's cracked, throw it away and try again with a new cap. If it's not, then use teflon tape or epoxy to screw the hose barb permanently into place.

9. ASSEMBLE THE LOWER PRESSURE TUBE
Glue together, in this order, the female-threaded PVC adapter, 1" length of ½" pipe, 90° slip elbow, 9" length of ½" pipe, and the end cap with the hose barb (Figure G).

10. ASSEMBLE THE LAUNCHER BASE
Pre-drilling as you go to prevent splitting the wood, attach the two 12" pine boards into a T

C

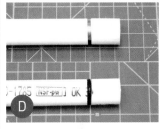
D

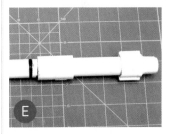
E

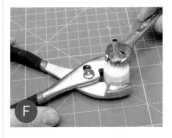
F

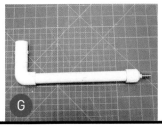
G

Mike & Patty Westerfield

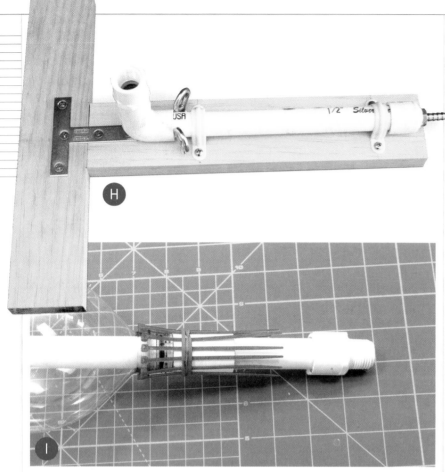

using the T-plate and wood screws. Next, fasten the lower pressure tube using the straps. Finally, place 2 large screw eyes into the base about 1" back from the vertical part of the tube (Figure **H**), creating a path for the launch string, which you'll add later.

11. POSITION THE ZIP TIES

Place a rubber band around the top of the slip coupling in the middle of the launch tube, and slide the bottle down until it rests on the coupling. Insert zip ties under the rubber band, leaving a little space between adjacent ties (Figure **I**), until you've gone all the way around. Secure the zip ties with a strip of duct tape and remove the rubber band.

12. TEST-FIT THE SLIP COLLAR

Slip the 1½" collar over the zip ties, adding or removing duct tape inside the collar until the tube stays in place over the ring of zip ties.

13. SECURE THE ZIP TIES

Near the bottom of the coupling, fasten the zip ties with the 1½" hose clamp (Figure **J**). This prevents the slip collar from sliding all the way down the launch tube when you release the rocket.

14. TRIM THE SCHRADER VALVE

Carefully carve away the valve's rubber coating with a hobby knife until you expose the inner brass lining (Figure **K**).

15. REMOVE THE VALVE MECHANISM

To allow air to safely escape, unscrew the middle pin with a pair of needlenose pliers (Figure **L**). If the pin is really stubborn, you can also drill it out from the bottom using a ³/₃₂" bit.

16. CONNECT THE HOSE

Soften both ends of the 20' length of ⁵/₁₆" clear vinyl tubing under hot running water, then slide it over the hose barb and Schrader valve and secure it with the small hose clamps (Figure **M**).

17. ADD THE LAUNCH STRING

Screw the launch tube into the lower pressure tube, then tie a string to one of the holes in the PVC collar. Thread the string through one screw eye, loop back through the second screw eye, and tie it off on the other hole in the PVC collar, forming a loop. Tie another string to the center of this loop, and cut this string to 22' or so — a bit longer than the vinyl tubing used to pressurize the rocket (Figure **N**).

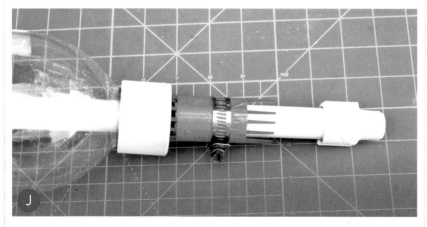

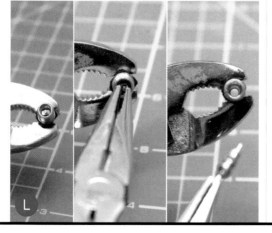

Pulling the string moves the PVC launch collar down, allowing the zip ties to spring apart and the rocket to launch.

BUILDING A WATER ROCKET

Now you need a rocket! Here's how to build a basic one. (For a parachute-recovered version, see makezine.com/go/soda-bottle-rocket.)

1. PREPARE THE BOTTLE (OPTIONAL)

Peel off the label if you wish, then clean it with WD-40. Remove any WD-40 with rubbing alcohol.

2. CUT AND MOUNT THE FINS

Cut 3 foamcore fins of pretty much any shape you like, just be sure the inside edge fits snugly against the base of the bottle (Figure **O**). You can find a template for the fins shown here at byteworks.us/Byte_Works/Make_Rockets.html.

Attach the fins using polyurethane glue to avoid damaging the plastic bottle. Then add a fillet of glue to strengthen each joint.

3. PAD YOUR ROCKET

Protect your rocket from crashes by putting some padding on top. I used a ring of pipe insulation (Figure **P**).

FLYING YOUR WATER ROCKET

Follow these steps to prepare and fly the rocket:
- » Drive 2 tent stakes into the ground and wedge the launcher's crossbar against them. This allows you to pull the launch string without dragging or tipping the launcher.
- » Fill the bottle about ⅓ full of water.
- » Insert the launcher into the rocket, then flip them over for launch.
- » Push the rocket down the launch tube so the heads of the zip ties are above the ring on the neck of the bottle. Slide the launch collar up over the edge of the zip ties to lock the rocket in place.
- » Move all spectators back at least 15 feet from the rocket and pressurize the system to 60psi–75psi.
- » If anything goes wrong, release the pressure before approaching the launcher. Built as directed, all you have to do is disconnect the air pump.

Once you're sure everything is ready, count down from 5 and tug on the launch cord to release the rocket. It will launch *very* quickly. Be ready to get wet — the spray will frequently give you a quick shower if the wind is blowing toward you! ◗

Share your build and trade launch tips at makezine.com/go/zip-tie-rocket-launcher.

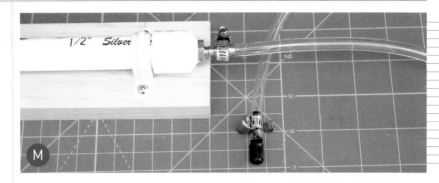

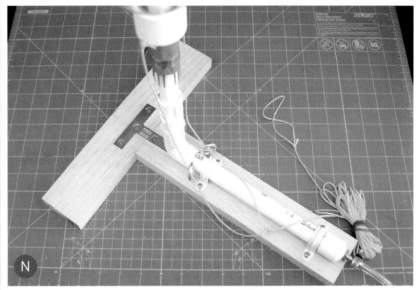

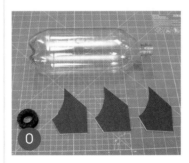

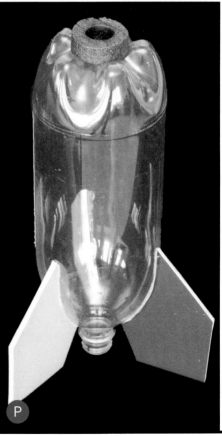

⚡ WARNING:

Water rockets can be dangerous. Never approach a pressurized rocket. The rocket is heavy when it lifts off, carrying over 1 pound of water, and it hits bone-breaking speeds of around 100mph in a fraction of a second. You can find the National Association of Rocketry's Water Rocket safety guidelines at waterrocketmanual.com/safety_code.htm.

Dark-Detecting LED Throwies

Written by Nick Normal

This simple circuit switches an LED on when the lights go out

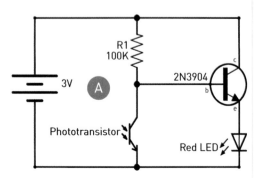

NICK NORMAL is a Queens-based artist and maker, and a lifelong biblioholic. He is a former five-year resident of Flux Factory, co-organizer for World Maker Faire (NYC), and advocate for all things geekathon.

MUCH LIKE OUTDOOR AND GARDEN LIGHTS, THE DARK-DETECTING LED THROWIE CIRCUIT switches on an LED when ambient light levels dip below a certain threshold. The big difference is the number of components: just five — a battery, phototransistor, resistor, LED, and transistor.

In this circuit (Figure **A**), the transistor and LED are effectively switched off when photons (particles of light) hit the phototransistor. When few or no photons hit the phototransistor, current freely passes through the transistor's collector-emitter junction, lighting up the LED.

1. LOAD AND SOLDER COMPONENTS

Begin with a standard through-hole circuit board design. We found that a small, round perf board comfortably contains all the components and fits a CR2032 battery holder perfectly. It also allows you to cleverly connect to the battery holder without any additional wiring.

Use a pair of helping hands to hold the PCB in place while you insert each of the components (Figure **B**) and solder them together using the schematic (Figure A) as your guide.

After soldering, trim the excess leads, except the outermost lead from the resistor (connected to the transistor's collector) and the negative leads from the phototransistor and LED. Bend these remaining leads through the nearest mounting hole (the larger holes) on the PCB (Figure **C**).

2. ADD THE BATTERY

Place the PCB on top of the battery holder. The leads from the PCB and the battery

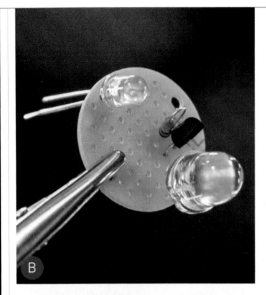

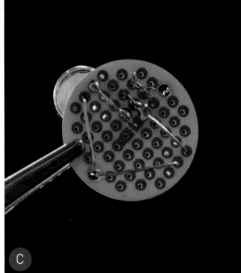

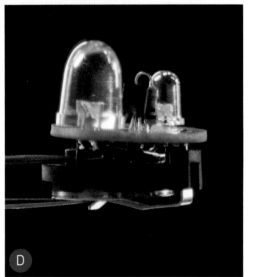

Time Required:
20-40 Minutes
Cost:
$5-$10

Materials
» **Phototransistor, infrared**
» **Transistor** 2N3904 NPN
» **Perf board, small**
» **LED, red jumbo, super-bright**
» **Resistor, 100kΩ, ¼W**
» **Thermoplastic pellets** such as Shapelock, Maker Shed #MKSHL1-500, makershed.com
» **Coin cell battery** CR2032
» **Battery holder** CR2032
» **Magnets**

Tools
» **Soldering iron**
» **Solder**
» **Wire stripper/cutter**
» **Helping hands with magnifier**
» **Cup or small bowl (glass or metal)**
» **Super glue**

CAUTION: Most consumer brands of thermoplastic pellets are safe to handle, but under no circumstance should you ingest them! I used a metal cup for my experiment, and just to be safe I now reserve that cup for this and other projects in my workshop.

holder's connectors should match up just enough for you to patiently solder them together (Figure **D**), with positive connected to the resistor lead and negative to the leads from the LED and phototransistor.

With everything soldered together, trim any excess leads and insert a CR2032 battery in the holder and voila, you have a Dark-Detecting LED circuit!

3. PREPARE THE THERMOPLASTIC
Now for some fun! I wasn't satisfied with the circuit simply living on a PCB, and wanted to encase it in something. I've been looking for an excuse to work with thermoplastic pellets, which turn soft and pliable in hot water and harden when cool. This was the perfect opportunity to experiment with the material.

All you do is pour boiling water over the pellets (ideally in a glass or metal container with a handle) and let them turn transparent, which takes around 15 seconds.

4. ENCASE THE CIRCUIT
When the pellets become clear, scoop them out with a spoon. They will immediately bond together and form a putty-like substance. Hot water will sometimes find its way into an air pocket, but otherwise the material is safe to mold with your hands. I dolloped it onto the PCB, and then began to shape it with my fingers.

Leave the phototransistor exposed, so it will still see light during the day. When the material completely dries it will harden and become mostly opaque white (Figure **E**).

5. ADD THE MAGNET
Finally, super-glue a magnet to the battery holder. Now toss your new Dark-Detecting

LED Throwie at something metallic. It will wait all day and then light up at night!

GOING FURTHER
The Dark-Detecting LED is a fun and simple circuit, but can easily be upgraded to the next level.

One obvious challenge is to ask yourself: "How do I operate this circuit using a white or blue LED?" These typically have a higher forward voltage than red LEDs, and will require some different components to activate the circuit. Or try connecting the circuit to a small solar panel and rechargeable power supply, so the batteries juice up during the day! Experiment and have fun. ●

Share your thoughts, tips, and mods on the project page at makezine.com/go/dark-detecting-led.

CHARLES PLATT is the author of *Make: Electronics,* an introductory guide for all ages, and its sequel *Make: More Electronics.* makershed.com/platt

**Time Required:
An Afternoon
Cost:
$5–$10**

Materials

» **XR2206 sound synthesizer chip**
» **LM386 amplifier chip**
» **7555 timer chip**
» **Capacitors, electrolytic:** 100µF (1), 33µF or 47µF (2)
» **Capacitors, ceramic:** 10nF (2), 4.7nF (2), 33nF (1), and 1µF (3)
» **Trimmer potentiometers:** 10kΩ (1), 20kΩ or 25kΩ (1), 500kΩ (2), and 1MΩ (1)
» **Resistors, ¼W:** 220Ω (1), 4.7kΩ (2), 5.6kΩ (3), 10kΩ (2), and 1MΩ (1)
» **SPDT switches, 0.1" spacing (7)** such as E-Switch #EG1218

Written by Charles Platt with Jeremy Frank

Wave Shaper

Build a multifunction sound synthesizer using a surprisingly fun audio chip

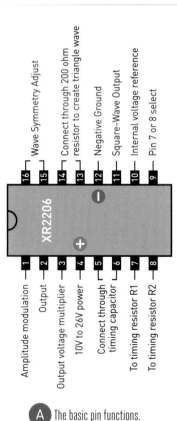

Pin labels (top, left to right):
16 — Wave Symmetry Adjust
15 — Connect through 200 ohm resistor to create triangle wave
14 —
13 —
12 — Negative Ground
11 — Square-Wave Output
10 — Internal voltage reference
9 — Pin 7 or 8 select

Pin labels (bottom, left to right):
1 — Amplitude modulation
2 — Output
3 — Output voltage multiplier
4 — 10V to 26V power
5 — Connect through timing capacitor
6 —
7 — To timing resistor R1
8 — To timing resistor R2

A The basic pin functions.

THE XR2206 IS A SLIGHTLY OBSCURE AUDIO CHIP THAT HAS SURPRISING POTENTIAL. It generates triangle waves and pure sine waves, and allows you to control them in unexpectedly imaginative ways. You can use it as the heart of a sound synthesizer.

I've had a lot of fun playing with the XR2206 since it was suggested to me by Jeremy Frank, who read my book *Make: Electronics* but now seems to know more about some topics than I do.

FREQUENCY AND AMPLIFICATION

The pin functions are shown in Figure **A**, and a basic test schematic is in Figure **B**. Don't be put off by the apparent complexity. Each section functions separately and is easy to understand.

For power, the XR2206 accepts 10VDC to 26VDC. You can find a simple 12V AC-to-DC adapter for less than $10 on eBay.

Pin 2 of the XR2206 is the audio output, which requires amplification. Pass it through a 33µF coupling capacitor and a 1M series resistor to the input of an LM386 single-chip amplifier, which has limited power, but is very simple (Figure **C**).

A resistor-capacitor combination sets the audio frequency — like a 555 timer, but simpler.

A timing capacitor goes between pins 5 and 6, and a resistance goes between pin 7 and negative ground. Use a capacitor value from 0.001µF to 100µF, and a resistance between 1K and 2M.

Frequency is calculated using a very simple formula: $f = 1,000 / (R \times C)$, where R is in kilohms and C is in microfarads. I chose a 33nF ceramic capacitor and 500K trimmer. Because I used the dual inputs on pins 7 and 8 (more on these later), the frequency is doubled, ranging from about 60Hz to 7kHz — close to the full audible spectrum.

The resistor network applied to pin 3 allows you to boost the chip's output. You can ignore pins 15 and 16, which enable you to fine-tune the wave symmetry, but aren't needed for our purposes. For now, leave inputs A, B, C, and D unconnected, put all the switches in their up positions, and let's get acquainted with the basic functions.

MAKING MUSIC

Switch 4, connecting pins 13 and 14 through a 220-ohm resistor, selects *sine-wave* or *triangle-wave* output. Sine waves have a natural quality — the kind of mellow tone you get by blowing across the top of a bottle. Adding higher frequencies to a basic sine wave can emulate various musical

Hep Svadja

Charles Platt

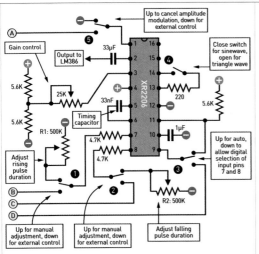

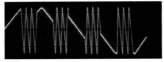

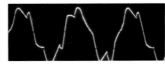

B A test schematic. You can leave inputs A, B, C, and D unconnected while testing the chip's main functions.

C Add this ultra-simple amplifier circuit to hear the sounds.

G A 7555 timer can create pulses and frequencies to control the XR2206 audio chip.

WAVE FORMS GALORE

D Move switch 5 down: the 7555 output makes the XR2206 alternate between 2 frequencies set by 500K trimmers.

E Switch 1 allows you to vary the amplitude very rapidly.

F Applying the 7555 timer output to pin 7 of the XR2206 chip produces unpredictable results.

instruments. A triangle wave (aka saw-tooth) is richer in harmonics, and sounds more artificial.

When switch 3 is up, the XR2206 automatically uses the resistances on pin 7 for the rising time and pin 8 for the falling time of each audio cycle. This allows you to create asymmetrical waves. You can view these signals with an oscilloscope; Figures **D**, **E**, and **F** show some that I created.

Now for some of the interesting features. Pin 9 is a digital input. In a logic-high state (2VDC or more) it selects the resistance attached to pin 7 to control frequency. In a logic-low state (1VDC or less) it selects the resistance attached to pin 8. If you feed a series of pulses to pin 9 — from a 555 timer, for instance — the XR2206 will flip between two different frequencies. If you do this very quickly, unique sounds are possible.

Because a 555 tends to create voltage spikes, I used an Intersil 7555, which has the same pin functions but a cleaner signal. If you've read *Make: Electronics*, the circuit shown in Figure **G** will look familiar as a basic frequency generator, with some additions. The 100uF capacitor stops frequencies from the XR2206 from trespassing into the 7555. The SPDT switch selects either a 4.7nF capacitor to generate low frequencies ranging from 150Hz to 10kHz, or a 1µF capacitor for 1Hz to 60Hz. The SPST switch rounds the

square-wave output with a 1µF bypass capacitor.

The complete breadboarded circuit is shown in Figure **H** (for a diagram of all component values, visit the project page online). I used E-Switch EG1218 SPDT slide switches, tiny and affordable.

Use any combination of switches 1, 3, 4, and 5 to apply the 7555 output to the various inputs of the XR2206. Run the 7555 slowly, and it can pulse the sound or create a tremolo or vibrato effect if you apply it to pin 1, which modulates the amplitude of the sound.

GOING FURTHER

Want more? You could substitute another XR2206 for the 7555 timer, so that your control pulses will have more subtlety than a simple square wave.

To create a keyboard-controlled synthesizer, open up a cheap keyboard and look for the contacts that close when keys are pressed. Each key must apply a different resistance to pin 7 of the XR2206, to generate a different note. You'll need a trimmer for each key; use an electronic guitar tuner to adjust each trimmer.

You can even convert the XR2206 into a tiny radio transmitter! Learn more about these hacks on the project page online. I have a feeling we've only just scratched the surface with this versatile audio generator. ●

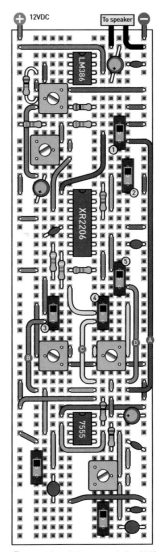

H Breadboarded circuit including 7555 control timer and LM386 amplifier.

See more photos, diagrams and XR2206 hacks: makezine. com/go/wave-shaper.

Ye Olde Brushless Motor

MICHAEL CURRY is a Kansas City, Missouri-based architect and designer whose projects range from sports stadiums to robotic petting zoo chickens. A co-founder of the MakerBot Design Studio, Michael has since created projects with Autodesk, Formlabs, Maker Media, Wiley Publishing, and others. skimbal.com

Time Required:
A Weekend
Cost:
$45–$70

Written by Michael Curry

Hep Svadja

Materials

» **3D-printed parts** Download free files for printing at thingiverse.com/thing:623901. Visit makezine.com/where-to-get-digital-fabrication-tool-access to find a machine or service you can use, or shop for great 3D printers at makershed.com.
» **U-bolts, #112 (4)** complete with nuts and metal brackets
» **Bearings, 608 type (2)** aka skateboard bearings
» **Bolts, M5, 10mm or 15mm (3)**
» **Wire, insulated, 22 gauge** Solid core is authentic; stranded or magnet wire also OK.
» **Electrical tape or heat-shrink tubing**
» **Arduino Uno microcontroller board** Maker Shed #MKSP99 or MKSP11, makershed.com
» **Solderless breadboard and jumper wires** Maker Shed #MKKN2 and #MKSEEED3
» **Transistors, MOSFET, N-channel (3)** SparkFun Electronics #10213, sparkfun.com
» **Battery, D cell**

Tools

» **Hacksaw and vise**
» **File or grinder**
» **Hobby knife**
» **Glue**
» **Computer with Arduino IDE software** free from arduino.cc/downloads
» **Project code** Arduino sketch *patent122944.ino*, free download from makezine.com/go/3d-printed-1872-motor

3D-print this 1872 replica Electro-Magnetic Engine that works like your drone's modern motors

FROM 1790 TO 1880 THE U.S. PATENT APPLICATION PROCESS REQUIRED THE SUBMISSION OF A WORKING "PATENT MODEL" to demonstrate the device being registered. These miniature machines were working showpieces created by skilled craftsmen. Today they offer insight into the intense technological advancement of the Industrial Revolution. Thousands have been collected by Alan Rothschild, proprietor of the Rothschild Petersen Patent Model Museum. Alan and I were thrilled to accept a challenge from Maker Media to create a 3D-printed working replica of one of the coolest models in the collection.

This Electro-Magnetic Engine, U.S. patent #122944, was patented in 1872 by Charles Gaume of Williamsburg, New York (Figure **A**). The model uses a system of rotating brass wheels to pulse electromagnet coils and rotate an armature of iron bars, converting electrical energy into mechanical. Today, this ingenious mechanism is duplicated electronically in the brushless motors used in everything from PC fans to quadcopter drones.

1. MAKE ELECTROMAGNETS

The 3 electromagnets are the heart of this motor model. You'll make each one from a U-bolt,

22-gauge wire, and your 3D-printed parts.

Trim the ends off 3 of the U-bolts with a hacksaw, using the 3D-printed cutting jig. (The fourth bolt is superfluous, but you'll need its bracket.) Clean up the cut ends with a file or bench grinder.

Clamp each trimmed U-bolt between the 2 halves of a printed magnet base, and secure with M5 bolts. Screw the nuts onto the U-bolts so they're flush with the ends.

Now you'll wind the electromagnet coils. The orientation of the wires is important: Start the first side by passing the wire behind the U-bolt (Figure **B**) and wrapping it around clockwise. Wind the wire tightly and neatly; you're going to be putting a lot of it on there.

Keep winding until you get to the nut, then wind a second layer, moving back down the bolt, always wrapping in the same direction (Figure **C**). Keep winding until your coil is 4 layers deep.

Bring the wire across to the other side of the U-bolt. Wind this side 4 layers deep too, except in the reverse direction: counterclockwise.

Wind the remaining 2 electromagnets the same way, then wrap the coils in electrical tape or heat-shrink tubing to keep them from unwinding (Figure **D**).

2. MAKE THE ROTOR

The rotor uses the 4 metal brackets from the U-bolts as the ferrous object that's acted upon by the electromagnets. Press each bracket into a slot on the rotor, then press the axle through the rotor to its stop, so the rotor is centered (Figure **E**).

3. ASSEMBLE THE FRAME

The rotor will spin on two 608 skateboard bearings. Press one bearing into each frame (Figure **F**).

Flip one frame over, so the bearing is facedown, and press the 3 electromagnets firmly into place (the holes are tight). Make sure to keep their orientations the same (Figure **G**).

Put the rotor in the bearing, then press the second frame into place. Finally, press the timing wheel and pulley onto either end of the axle (it doesn't matter which; they're ornamental).

4. WIRE THE MAGNETS

Connect the 3 right-hand wires together to create one common positive lead. You'll connect this to the positive side of the Arduino. The other 3 leads are the negative leads, which connect the electro-magnets to their switching MOSFETs; I like to add colored wire to the ends to indicate which is which.

To make an old-fashioned-feeling cloth power cable, run all 4 leads through a wide shoelace.

5. HOOK UP THE ARDUINO CONTROLLER

The 1872 patent model used a mechanical system to pulse the coils in the right sequence to turn the motor. I replaced it with an Arduino and MOSFET transistors — easier to build and to tinker with.

The electromagnets need 1.5V and lots of current, both drawn from a D battery. The Arduino sequences the MOSFETs, which switch the coils on and off. Breadboard the circuit (Figure **H**), following the diagram on the project page online. Then upload the project code to your Arduino.

6. START YOUR MOTOR.

Flick the motor lightly with your fingers to get it spinning. (The Arduino sequencer code is very simple, and doesn't have a way to accelerate the motor from a dead standstill.) Now you've got a working replica 1872 electromagnetic motor! ●

For the breadboard diagram, more step-by-step photos, videos, finishing tips, and the project code, visit makezine.com/go/3d-printed-1872-motor.

This project is excerpted from *Inventing a Better Mousetrap: 200 Years of American History in the Amazing World of Patent Models* by Alan Rothschild and Ann Rothschild (Maker Media), available at the Maker Shed (makershed.com) and fine bookstores.

Michael Curry

3 Fun Things to
3D Print
Written by Eric Chu

1. WIND-UP SPRING MOTOR
by Greg Zumwalt
thingiverse.com/thing:402412
Fully 3D-printed, this wind-up motor uses the flexibility of the PLA plastic to create a working spring. Car-chassis and wheelie-dragster versions (thing:430050 and thing:452248) put the stored energy to good use!

2. TINY VACUUM CLEANER
by Michael Curry
thingiverse.com/thing:539986
Need a small desktop vac? Print this one and learn how a vacuum cleaner works while you're at it. It's driven by a DC motor, and a piece of bathroom tissue acts as the filter.

3. BUILD YOUR OWN JET ENGINE
by Patrick Saville, General Electric
thingiverse.com/thing:392115
This crank-operated, cutaway model jet engine is fully 3D printed and assembled with glue. And check out the nifty 2D printable box, to package the engine as a gift or for display.

Written by
Forrest M. Mims III

How to
Connect Optical Fibers to LEDs and Sensors

FORREST M. MIMS III
(forrestmims.org), an amateur scientist and Rolex Award winner, was named by *Discover* magazine as one of the "50 Best Brains in Science." His books have sold more than 7 million copies.

Skip the store-bought couplings and make these easy DIY fiber hookups

Time Required:
15–30 Minutes

Cost:
$10–$20

Materials
» **Plastic optical fiber, 2.2mm**
» **LEDs, phototransistors, or photodiodes, 5mm epoxy encapsulated**
» **Heat-shrink tubing, 6mm or 8mm**
» **Plastic cement or cyanoacrylate (CA) glue** aka super glue

Tools
» **Hobby knife**
» **Sandpaper: 200 grit and 400 or 600 grit**
» **Magnifier, 10x**
» **Drill or high-speed rotary tool** Dremel 7700 or similar
» **Drill bits: 3/64" and 7/64"**
» **Vise or clamp**
» **Heat gun or butane lighter** or other source of heat

OPTICAL FIBERS ARE ULTRA-CLEAR STRANDS OF PLASTIC, GLASS, OR SILICA CONSISTING of a central core surrounded by a cladding and a protective coating. Light injected into the core of a fiber remains trapped until emerging from the opposite end. This lets you transmit light point-to-point with very little loss, and even bend it around corners. The light stays in the core because the cladding has a slightly higher index of refraction than the core.

Silica optical fibers are primarily used to transmit high-bandwidth data over long distances. Inexpensive plastic fibers are widely used in sensors, illuminators, and toys. They're also used to couple light to photodiodes in environments that require electrical isolation or protection from the elements or a corrosive environment. Plastic fibers are also used to illuminate displays and to send light through openings too small for a flashlight.

CONNECTING FIBERS TO LEDS AND SENSORS
Optical fiber couplers for various LEDs and light sensors are commercially available, but you can skip the connector and simply connect silica and plastic fibers directly to LEDs and sensors. For the examples described here, I used LEDs encapsulated in standard 5mm clear epoxy packages, and 2.2mm-diameter plastic fiber having a 1mm core and a black polyethylene jacket. I used fiber from Jameco Electronics (jameco.com) but many kinds of fiber are available from other online sources, including eBay, and the methods described here can be adapted for use with most of these fibers.

PREPARING THE FIBER ENDS
For most applications, best results are obtained with a fiber having flat, smoothly cut or polished ends. A simple way to achieve this with plastic fiber is to place

Hep Svadja

the end of the fiber on a wood surface and slice off a few millimeters by pressing a sharp hobby knife blade straight down into the fiber. Press the cut end of the fiber against 200-grit sandpaper on a flat surface and rotate the fiber in a dozen or so circles across the paper. Follow this with 400 or 600 grit paper. A fiber end polished in this fashion is shown in Figure A. The roughened edges of the fiber's jacket can be trimmed away with a hobby knife.

HEAT-SHRINK TUBE CONNECTION

Heat-shrink tubing provides the simplest way to connect optical fibers to LEDs and sensors. This method is not necessarily practical for long-term use, especially outdoors, but it works well for basic experiments and demonstrations with 2.2mm fiber. For best results, use 6mm- or 8mm-diameter heat-shrink tubing and a 5mm LED or sensor.

Slip a 1" length of tubing over the LED or sensor and warm the tubing with a heat gun until it holds the LED or sensor tightly in place. Slip the end of the fiber into the open end of the tubing and continue warming. Depending on the diameter of the tubing, the fiber will either be anchored in place or it can be slipped in and out of the tubing (Figure B). You can keep it in place by applying some adhesive or you can allow the fiber to be removable.

PERMANENT DIRECT CONNECTION

A stronger coupling can be made by cementing the fiber into a hole bored into the end of the LED or sensor's epoxy capsule. You can cement the jacketed end of the fiber, or remove a section of jacket and cement only the fiber itself. An inch or so of heat-shrink over the junction will finish the job. Follow these steps:

TIP: If you're using the fiber with a light sensor, paint the exposed base of the sensor with black enamel to block external light. Additional blockage may be needed, as infrared wavelengths may penetrate black paint.

1. Secure the leads of the LED or sensor in a vise or a DIY clamp made from a clothespin and a large binder clip (Figure C).

2. Use a fine Sharpie to draw a plus sign (+) directly over the end of the LED. Insert a ³⁄₆₄" bit into the chuck of a drill. I prefer the handheld, battery-powered Dremel 7700.

Lightly touch the spinning bit to the center of the plus sign. Let the drill do the work as you guide it straight into the device while applying very gentle pressure (Figure D). Carefully bore the hole to just above the tiny wire(s) that make contact with the light-emitting or -sensing chip (Figure E).

The hole produced by a ³⁄₆₄" bit should accept the 1mm bare fiber core. If you're connecting jacketed 2.2mm fiber, carefully enlarge the hole with a ⁷⁄₆₄" bit.

3. Use compressed air to blow away any chips in the hole. Be sure any connection leads on the top of the chip are undamaged.

4. Insert the polished end of the fiber into the hole, then secure it with cyanoacrylate or other plastic adhesive and let it dry. Figure F shows a fiber with a bare end inserted into a blue LED, and Figure G a jacketed fiber inserted into a white LED.

5. Insert a suitable length of dark heat-shrink tubing over the LED or sensor and 1" or so of the fiber, and warm it to secure the tubing in place.

GOING FURTHER

You can start using your connected fibers right away for illuminating your projects, props, and models, or for photography — I had fun experimenting with "light painting" by making 10-second time exposures (Figure H). Pulse the LED to create dashed instead of continuous lines in the images, and use multiple fibers to add more colors.

In a future column I'll show you how to use fibers connected to LEDs and phototransistors as sensitive sensors.

Meanwhile, you can expand on the methods described here by designing your own connections. Use a ballpoint pen housing to make a handheld optical fiber pixel probe or micro-light source. Or consider 3D printing your own custom-designed connectors and fixtures. ◐

How will you use your DIY fiber optics? Share ideas and projects at makezine. com/go/DIY-fiber-optic-connections

Forrest M. Mims III & Hep Svadja

A

B

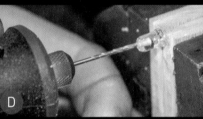

C

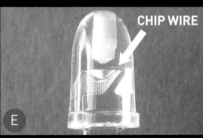

D

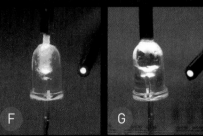

CHIP WIRE

E

F

G

H

1 2 3 Cosmic Ray Detector

Written by Kranti Gunthoti ■ Illustrated by Andrew J. Nilsen

alpha

muons

electrons

BUILD THIS SIMPLE CLOUD CHAMBER AT HOME AND DETECT cosmic ray muons, electrons, and alpha particles.

1. SECURE FELT
Sandwich a section of felt inside the bottom of the cup between two magnets — one inside the cup, the other on the outside.

2. PREPARE CUP
Wearing safety glasses, pour in just enough isopropyl alcohol to completely soak the felt.
» Form a Play-Doh seal about ½" thick on the cup rim to avoid any alcohol vapor leakage between the cup and metal plate.

3. STICK CUP ON PLATE
Wearing an insulated apron and gloves, put some dry ice into the styrofoam cooler lid. Place the metal plate on top and wait until the surface becomes very cold.
» Invert the cup on the metal plate and press firmly so that the cup sticks to it. If your Play-Doh becomes too dry after coming in contact with isopropyl alcohol, try substituting plasticine.

With the lights off, illuminate the area closest to the plate with your flashlight. After 2-3 minutes the alcohol vapor will cloud into a supersaturated state. When an electrically charged particle zips through, it will ionize the vapor atoms by ripping off the electrons along its path. This will trigger the condensation process and we can see a trail from the charged particle. ◎

KRANTI GUNTHOTI & SAMATHA KRANTHIJANYA are passionate about physics. One of their favorite pastimes is designing science experiments to explore how nature works.

Time Required:
1-2 Hours
Cost:
$20–$40

You will need:
» **Isopropyl alcohol, 99%**
» **Dry ice (solid CO_2)**
» **Felt, black**
» **Cup, clear plastic**
» **Play-Doh or plasticine**
» **Plate or pan, black metal**
» **Flashlight**
» **Magnets, neodymium (2)**
» **Cooler lid, styrofoam**
» **Gloves, insulated**
» **Safety glasses**
» **Apron, insulated**

Different particle trail types:
» Thin and long tracks are **muons**, originating from cosmic rays

» Thick, small tracks are **alpha** particles, caused by radon atoms in the atmosphere — not related to cosmic rays

» Wiggly tracks are low-energy **electrons** experiencing multiple scattering

To watch the cloud chamber in action visit makezine.com/go/simple-cloud-chamber

GIVE A GIFT.
GET A GIFT.

Make:

Give your favorite geek **permission to play**. Send a year of MAKE (6 bimonthly issues)for **just $34.95**, a savings of 42% off the cover price.

YOUR FREE GIFT
2015 Guide to 3D Printing PDF

GIFT FROM

NAME
_____

ADDRESS
_____

CITY   STATE   ZIP
_____

COUNTRY
_____

EMAIL ADDRESS *required for delivery of the Ultimate Guide to 3D Printing PDF

GIFT TO

NAME
_____

ADDRESS
_____

CITY   STATE   ZIP
_____

COUNTRY
_____

EMAIL ADDRESS *required for access to the digital editions

We'll send a card announcing your gift. Your recipients can also choose to receive the digital edition at no extra cost. Price is for U.S. only. For Canada, add $5 per subscription. For orders outside the U.S. and Canada, add $15.

PROMO CODE 454GS1

BUSINESS REPLY MAIL
FIRST-CLASS MAIL PERMIT NO. 865 NORTH HOLLYWOOD, CA

POSTAGE WILL BE PAID BY ADDRESSEE

Make:

PO BOX 17046
NORTH HOLLYWOOD CA 91615-9186

JET PACK!

WRITTEN BY WARREN SIMONS HOWTOONS

IT WAS LATE OCTOBER WHEN THE **SHORTWAVE RADIO** LIT UP LIKE A **JACK O' LANTERN.**

I WAS WORKING ON A **TOP-SECRET PROJECT**, AND ALMOST DIDN'T HEAR THE GRAVELLY VOICE ECHOING THROUGHOUT THE LAB...

ORANGE STAR... --KRRSHH-- DO YOU READ ME...OVER...

IT WASN'T A SURPRISE, THOUGH. THERE WERE WHISPERS IN SCHOOL THAT SOMETHING BIG WAS HAPPENING ON **OCTOBER 31ST...**

ORANGE STAR... --KRRSHH--DO YOU --CRACKLE--COPY...

SOMETHING THAT WOULD AFFECT **ALL** OF US...

WE HAVE AN URGENT MISSION FOR ORANGE STAR... --KRRSHH-- AM I GETTING THROUGH--

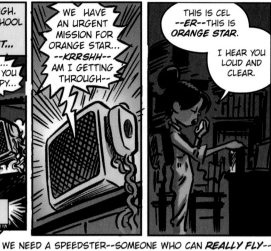

THIS IS CEL --ER--THIS IS **ORANGE STAR.**

I HEAR YOU LOUD AND CLEAR.

ORANGE STAR! WHEW! IT'S A RELIEF TO HEAR YOUR VOICE!

I REALIZE THAT IT'S SHORT NOTICE, BUT THE **SUPER SECRET SCIENCE COUNCIL** NEEDS YOUR HELP!

HALLOWEEN IS ONLY A FEW DAYS AWAY, AND WE THINK ONE OF **DR. MANIACLE'S** SUPER-MEAN ROBOTS IS GOING TO TRY AND STEAL ALL OF THE CANDY FROM THE CITY!

WE NEED A SPEEDSTER--SOMEONE WHO CAN **REALLY FLY**-- TO COLLECT AS MUCH CANDY AS POSSIBLE! CAN YOU DO IT, **ORANGE STAR?**

YOU'VE GOT THE RIGHT GIRL FOR THE JOB, COMMISSIONER.

I'M ON IT!

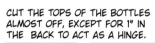

USE A PAIR OF 2-LITER BOTTLES TO CREATE A **JET PACK!** A PAPER TOWEL ROLL MAKES A GREAT SPACER.

MAKE CONCAVE CUTS TO MATCH BOTTLE SURFACE.

GLUE SPACERS, CONNECTING BOTTLES.

CUT TWO SLITS AT THE TOP AND BOTTOM OF BOTTLES. WEAVE A BELT THROUGH FOR STRAPS.

MAKE EXHAUST PIPES! CUT OUT BOTTOMS OF PAPER CUPS, AND GLUE TO THE BOTTLE BOTTOMS.

CREATE FLAMES BY SHREDDING STRIPS OF RED AND YELLOW PAPER!

CUT THE TOPS OF THE BOTTLES ALMOST OFF, EXCEPT FOR 1" IN THE BACK TO ACT AS A HINGE.

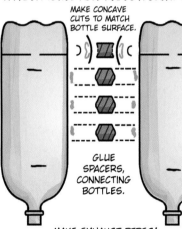

USE TAPE AS A LATCH.

FILL BOTTLES WITH CANDY!

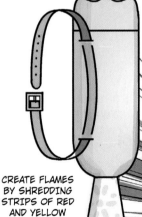

NO WAY ROBOTS ARE THREATENING **MY** CITY!

TONIGHT, I'M GOING TO COLLECT THE MOST CANDY IN THE **ENTIRE** NEIGHBORHOOD!

CELINE'S READY FOR HALLOWEEN... ARE YOU?!

Willebrord Snell and Triangulation

Written by William Gurstelle ■ Illustration by John Thomas

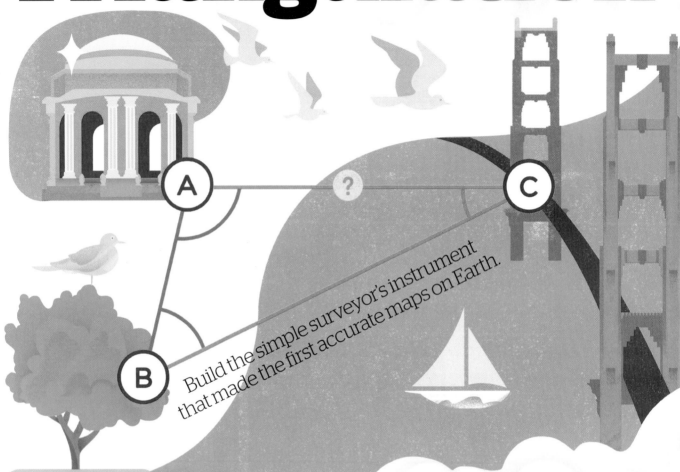

A

?

C

B

Build the simple surveyor's instrument that made the first accurate maps on Earth.

Time Required:
30-60 Minutes

Cost:
$35–$40

Materials
» **Angle finder, plastic** Such as Harbor Freight #94963
» **Carpet tacks or small box nails (2)**
» **Small block of plastic, 1"×1"×½"** You can glue two ¼"-thick pieces together if needed.
» **Camera tripod** Cheap ones are $15–$20.

Tools
» **Drill and ³⁄₁₆" bit**
» **Hole tap, ¼"-20 UNC, with handle**
» **Glue**

THE BURGHERS OF LEIDEN, HOLLAND, MUST HAVE WONDERED WHAT THAT ODDBALL FROM THE UNIVERSITY WAS UP TO. Throughout the years 1614 and 1615, they saw Willebrord Snell, a young professor of mathematics, repeatedly climbing up and down the town's church steeples and bell towers, lugging a huge quarter-circle of iron. Then they saw him carefully rolling out and then rolling up a long metal chain, all the while carefully writing down notes in a notebook. What could this fellow be up to?

It turns out he was busy making scientific history. Professor Snell was inventing the science of geodesy and laying the baselines for the future practice of surveying.

Specifically, Snell was attempting to figure out the exact size of the Earth. The first attempt was made nearly two millennia earlier by the Greek scientist Eratosthenes of Cyrene, who used the different lengths of the sun's shadows at noon in two towns as the basis of his calculations. The circumference that Eratosthenes came up with was pretty accurate, but Snell thought he could do better with a new technique he'd invented, using triangulation for the purpose of measuring very long distances. Snell's method was a game changer, and soon, people of science all over Europe saw what an incredibly valuable tool triangulation was.

Prior to Snell's work, the only way to reckon the distance between two towns or landmarks was to measure it directly. A common method was to make a couple of extremely long rulers and leapfrog them along the flattest and most direct path possible between the two points. That got old fast, so somebody came up with the slightly better idea of counting the turns of a wagon wheel and then applying basic algebra to figure out how many feet or yards the wagon had traveled. Neither of these methods was very accurate, and if there was something in the way of the end points, say a mountain or a river, these methods didn't work at all.

As a math professor, Snell knew well the trigonometric principles formulated by early Greek and Arab mathematicians. From their work, he understood that every triangle made of three sides and three angles, and further, that if you know the values of two angles and one side, or two sides and one angle, then you can accurately calculate all the other sides and angles.

Specifically, Snell used two trigonometry formulae — the law of sines and the law of cosines — to invent the practice of modern triangulation: the calculation of the distance between points by first making an accurately measured baseline and then measuring the angles made between the distant point and the two ends of the baseline. Snell's great innovation was to accurately calculate new baselines from previous measurements instead of physically measuring them all, revolutionizing the speed and accuracy of surveying large areas of land. With this technique and his trusty quadrant (the iron gadget that mystified the burghers) he charted all of Holland. Soon after, surveyors and mapmakers were redrawing the world more precisely than ever before.

UNDERSTANDING TRIANGULATION

Triangulation isn't a difficult technique to grasp. In this edition of Remaking History, you'll make a simple piece of measuring equipment that allows anyone to calculate distances for making good maps.

Suppose you need to determine the distance from point Ⓐ (the Palace of the Fine Arts) to point Ⓒ (the Golden Gate Bridge) as shown in the diagram on the previous page. Because the bay is in your way, you can't measure the distance directly, although you can see point Ⓒ from both point Ⓐ and point Ⓑ (a nearby tree).

First, make an accurately measured baseline between points Ⓐ and Ⓑ. The longer your baseline is, the more accurate your subsequent distance calculations will be. Use a reel-type tape measure or other direct measuring device for this.

Then, go to point Ⓐ and measure the angle between point Ⓑ and point Ⓒ. Go to point Ⓑ and measure the angle between points Ⓐ and Ⓒ.

You now know two angles. Use the fact that the three angles in a triangle must add up to 180° to calculate the third angle.

Now you've got all the information you need to calculate the length of segment AC. You'll simply plug the quantities into your calculator and use the Law of Sines:

$$\frac{\sin CAB}{BC} = \frac{\sin ABC}{AC} = \frac{\sin ACB}{AB}$$

You know the length of segment AB, and the values for angle ACB and angle ABC. Just look up the sine values, set up the rate pairs, and solve for AC.

BUILD YOUR DIY SURVEYOR'S SEMICIRCLE

The key to accurate triangulation is measuring angles precisely. To do that, you'll make a Snell-style angle-measuring instrument: a surveyor's semicircle, also known as a graphometer or semicircumferentor, the ancestor of today's theodolite.

1. Drill a ³⁄₁₆" hole in the center face of the 1"×1"×½" plastic block.

2. Cut screw threads in the hole using the ¼"-20 UNC tap.

3. Glue the block to the bottom of the angle finder.

4. Glue the sighting nails to the knobs on top of the angle finder, as perpendicular as possible.

5. Screw the tripod's mounting screw into the tapped hole.

To find angles between distant objects, use the nails as sights, lining up the object with both the front and back nail. Note the angle separating the objects on the dial. ◉

WILLIAM GURSTELLE is a contributing editor of *Make:* magazine. His latest book, *Defending Your Castle*, is available at all fine bookstores.

Samuel DeRose

+SKILL BUILDER
HOW TO CUT THREADS

The tool you use to cut a male threaded piece is called a *die*. When you make a female threaded piece, you use a *tap*. In this case you're making threads in a hole, and that means you're making a female threaded piece.

A. Start the threading process by carefully positioning the main axis of the tap parallel to the hole.
B. Turn the tap ½ turn, and then back out the tap ¼ turn to remove shavings so the tap doesn't get clogged.
C. Just keep doing this until the hole is fully threaded!

+See step-by-step photos at makezine.com/go/tap-threads.

For complete examples of how to measure distances using your graphometer, or even easier, to measure the height of a flying drone or other object, visit the project page at makezine.com/go/diy-surveyors-semicircle.

1 2 3 Grow Your Own Crystals

Written and Photographed by Emma Chapman

PART CRAFT AND PART KITCHEN EXPERIMENT, GROWING YOUR OWN ALUM CRYSTALS IS EASY to do at home. I used eggshells for a geode-like effect, but you can experiment with growing them other ways, such as on rocks or suspended fishing line.

1. SLICE THE SHELLS

Use an X-Acto knife to cut your eggshells in half lengthwise. This is messy; do it over the sink. Then wash out the eggshells and let them dry.

2. ADD SEED CRYSTALS

Paint a thin layer of Elmer's glue onto the inside of the eggshells, then sprinkle alum over them and let dry overnight. These are your seed crystals. They'll give the dissolved alum (Step 3) someplace to attach to and start growing.

3. MIX AND WAIT

Heat 2 cups of water until it's just about to boil. Add the food dye (about 40 drops of dye per color) and ¾ cup alum, then stir until the alum is completely dissolved.

Allow the mixture to cool for 10 minutes. Pour the mixture into your container and, wearing gloves to avoid staining your hands, add one eggshell, open side facing up.

Let your crystals grow for 8+ hours in this dye bath. I let mine grow for about 20 hours, and the crystals got quite big! Finally, rinse off your finished crystals and allow them to dry. ◕

EMMA CHAPMAN and her sister Elise are the co-founders of women's lifestyle company abeautifulmess. com. They share home decor projects, recipes, and crafts, as well as bits of their lives in the Midwest.

**Time Required: 1–2 Hours
Cost: $5–$10**

You will need:

» **Eggs**
» **Alum** sold in the spice aisle. I do not recommend powdered alum. For best results, use the granulated type of alum used in pickling — the granules look like tiny crystals (because they are).
» **Elmer's glue**
» **Food dye**
» **Hot water**
» **X-Acto knife**
» **Paint brush**
» **Plastic containers** that you don't mind staining
» **Gloves**

See more photos and share your crystal crafts at makezine.com/go/123-crystals.

Toy Inventor's Notebook

FOAM PLATE FALCON

Invented and drawn by Bob Knetzger

Time Required:
A Few Minutes
Cost:
$0–$1

Materials

» **Foam picnic plate**
» **Penny**
» **Tape**
» **Scissors**
» **Computer and printer**
» **Hobby knife or hot-wire foam cutter**

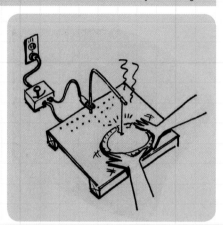

TIP: You can cut the parts with a hobby knife, but for extra fun, why not use your "Five Minute Foam Factory" (FMFF) hot wire cutter from *Make:* Volume 16 (makezine.com/go/5-minute-foam-factory). With it you can slice out multiple flying falcons from a stack of plates all at once!

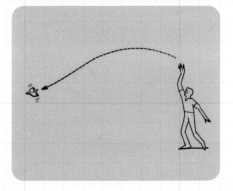

QUICK — MAKE THIS FUN MINI-PROJECT BY UPCYCLING A FOAM PICNIC PLATE INTO A FLYING FALCON!

Go online to makezine.com/go/foam-plate-falcon to download the full-size pattern. Print it on thick paper or cardstock and carefully cut out the 2 shapes, one for the wings/body and one for the tail fin. Be sure to cut out the thin slots for the tail and for the ailerons.

If you're cutting by hand, place the pattern on the plate and trace all around with a ballpoint pen. Trace the tail, too. Then cut out carefully with a hobby knife or scissors.

If you're cutting with the FMFF (see tip, above), pin the pattern in place onto a foam plate and connect the power to the cutting wire with the alligator clips. Adjust the voltage so that the wire is just hot enough to slice the foam, but not hot enough to

smoke or burn. That will minimize harmful fumes — but always cut foam outside or with plenty of ventilation! The wire slices though the foam easily as you slide it along the edge of the pattern. If you go off course, just go back and trim again. Be sure to cut the slots, too.

To assemble, slide the tail into the slot as shown. Add a penny with tape for a nose weight. Bend the ailerons up a bit and adjust the tail to be perpendicular. The curved shape of the plate makes a cool dihedral angle on the wingtips.

To fly, toss gently overhand from a high point. The falcon should glide down. Adjust the ailerons and position of the weight if needed. Enjoy! ◆

Download the pattern and show us your foam falcon flights at makezine.com/go/foam-plate-falcon.

Cordless Precision Screwdriver

$20 : craftsman.com

OK, it's not *that* hard to use a set of standard precision screwdrivers when working on a project. But with a growing stack of broken mini-drones needing repair along with many other small electronics projects, the idea of hours of twisting tiny screwdrivers with my fingertips began to overwhelm me to the point of ignoring everything entirely.

Then I found this powered precision screwdriver. With the push or pull of its switch, I'm able to zip all sorts of minuscule screws in and out of their threaded homes. Its quick speed outpaces my manual capabilities, and the torque has proven to be high enough to yank out very tightly fit screws. Even the most ambitious laptop disassembly feels like a breeze.

The driver comes with six precision bits that fit most of the screws I encounter — two sizes each of Phillips, slotted, and Torx — and it uses a standard 4 mm (⁵⁄₃₂") bit size, which means I can swap in more exotic bits from a bigger set I already own. The LED light at the tip finally gives me a clear indication if I have the bit properly attached to the screw as well.

There are a few handicaps, although none are deal-breakers. Using the switch takes a little getting used to — you have to hold the driver like a pencil rather than like a screwdriver. The plastic casing doesn't feel extremely heavy duty, although that does help keep the weight down. And the mechanism's body blocks access to deep-set screws.

Altogether, I'm very pleased, and the wake of newly disassembled electronics around my house proves it. *— Mike Senese*

Hep Svadja

SOLDERING AID KIT
750 ASSORTMENT
$290 : pcbgrip.com

The PCB Grip electronics assembly system has an answer to every circuit board work-holding problem I could come up with — from simply holding the boards, to holding down components, to clamping loose wires and components for soldering to each other. They even offer probe clamps that can fit test leads or probes, and goosenecks for creating flex-arm "third hands."

I was pleasantly surprised at how adjustable the system was, and how well the components were made. They thought of everything, even including thumbscrews for most, if not all, of the adjustments that you might want to make.

You can buy individual components as needed, or one of several versatile kits. An added benefit is that it makes use of open source aluminum extrusions and standard hardware, which makes it especially expandable.

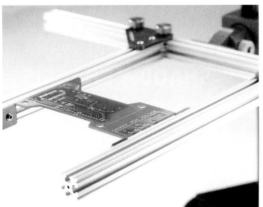

If you have ever struggled with getting your electronics components to behave during soldering or testing, this is the system to try.

— *Stuart Deutsch*

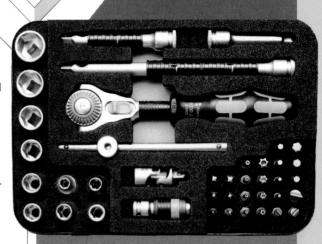

WERA ZYKLOP
SPEED SOCKET SET
$300 for 38-piece, ³/₈"
www-us.wera.de

Sometimes I take my projects to my tools, other times I take my tools to my projects. Wera Zyklop Speed tool sets come with a swivel-head ratchet, useful drive accessories, and sockets — often including sizes that most other sets skip over (such as ¹¹/₃₂"). Larger sets also include a bit adapter and screwdriver bits so that you can work on a myriad of screws and not just hex nuts and bolts. These sets are small and compact, but they contain everything I need to tighten or loosen most fasteners for whatever robotics, electronics, and DIY projects I work on.

What I like most is that the Zyklop Speed swivel-head ratchet can be used at different angles, from 0° for fast, screwdriver-like spinning to 90° for higher-torque tightening and loosening. I also appreciate that they include smaller, Maker-friendly socket sizes that you won't find in automotive socket sets.

These sets are available in ¼", ⅜", and ½" drive sizes, and with your choice of imperial or metric configurations. The larger sets come with locking extensions, wobble extensions, universal joints, sliding handles, and bit holders, so that you don't have to cobble together accessories from myriad different sources.   — *SD*

POCKET AUTOMATIC
CENTER PUNCH
$16.55 : generaltool.com

When you want to drill a hole in metal, or other hard or slick materials, the bit can wander off the mark. For precision drilling, a center punch is essential — it leaves a sharp dimple that helps guide your drill bit. Traditional manual punches require two hands, one to hold the punch and another to use a hammer. With your third hand, hold the workpiece steady. You don't have a third hand? Get an automatic center punch.

I use this General Tools punch all the time. There are more expensive models out there, and I've heard some criticism of this one, but I love it. The spring is adjustable for working with different materials; I've used mine for everything from PVC plastic to mild steel. I have even tested it on the butt end of a high speed steel drill bit, with tension set to its highest setting, although using it on very hard materials will likely dull the tip very quickly.

The narrow point on this punch doubles as a scribing tool for marking metal before cutting. It is one of the most frequently used hand tools in my shop. Get one and you won't regret it.

— *Andrew Terranova*

TOOLBOX

BUILDING OPEN SOURCE HARDWARE

By Alicia Gibb
$29 : amazon.com

Building Open Source Hardware, edited by Alicia Gibb, is the single best introduction to open source hardware available. Gibb is a long-time proponent of open hardware; she shaped the very definition used by the Open Source Hardware Association and continues to champion the ideal that sharing your hardware designs with the world is a good thing. Even if you are already familiar with Open Hardware concepts and vision, you will gain new insights from many of the leading minds in the community. Perhaps the best part of the book is Chapter 6, where Gibb shares her hardware design for Blinky Building, an outline of the Empire State Building adorned with LEDs, and instructs the reader how to create their own derivative of the project. This book is not to be missed.

— *David Scheltema*

NEW MAKER TECH

VEX ROBOTICS GIANT HEXBUGS

$50-$90 : hexbug.com/vex

When it comes to interesting and unique robots you can buy for under $100, there typically aren't many options. However, collaboration between VEX and Hexbug has yielded some fantastic results ranging from $50 to $90.

Four models are available: the peculiarly wheeled Ant bot, the Spider and Scarab hexapods, and the rather large Strandbeast with an interesting 8-legged mechanism. Assembly is fairly simple, though it can be somewhat time-consuming. Touted as being able to switch from radio-control to autonomous mode, our tests found the autonomy lacking, but radio-controlled mode to be quite fun. If you're looking for an entertaining kit to steer around the house, these are a great alternative to a radio-controlled car.

— *Caleb Kraft*

COLORED BREAKAWAY HEADERS

$0.99 each, price break with quantities of 5+ : pololu.com

If you have ever used breakaway headers before, I don't need to tell you how useful they are. They allow breakout boards to attach to breadboards, shields to attach to Arduinos and other microcontroller and mini computer systems, and offer a means by which to quickly attach components to a soldered circuit board or temporary breadboard circuit.

Now Pololu has increased their utility even further, releasing colored breakaway headers in 1×40-pin male-male strips with 0.100" spacing, in 4 new colors: red, yellow, blue, and white.

Here are some of the ideas I have for the first batch of headers I ordered: red and black for V+ and ground power rails; red, yellow, and black for a row of servo connections; blue, white, and yellow to differentiate sensor input pins in a robotics project. You get the picture.

— *SD*

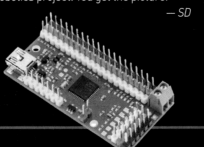

SHADOW CHASSIS

$13 :
sparkfun.com

What sort of robot chassis would you expect for thirteen bucks? With the Shadow Chassis you get a reasonably sturdy plastic frame that snaps together without tools, taking advantage of precisely machined parts to fit together with nary a wobble. It's a great beginner's entry into robotics, and would be a good starting point for a high school robotics class learning platform.

Adding parts is a cinch: Take advantage of the variety of mounting holes on the top panel to attach your own hardware — including a servo mounting hole and some Actobotics-compatible mounting holes — or simply drill into the ABS to create your own solution. Heck, it's just $13. If your customizations don't work for a future project, start fresh with a new chassis. SparkFun also offers an $85 "basic kit" (P/N 13166) that includes snap-in 140-RPM gearmotors, a 4xAA battery pack, wheel encoders, a microcontroller, and a couple of sensors.

— *John Baichtal*

GETTING STARTED WITH DRONES
By Terry Kilby and Belinda Kilby : $25

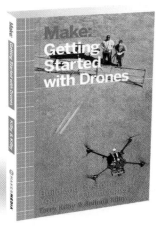

Perhaps the most misunderstood object in today's airspace, the drone is an incredibly useful hobbyist's tool that is as much fun to build as it is to fly. With an emphasis on safety and responsible use, *Getting Started with Drones* begins with the history of this tiny flyer and takes the reader step-by-step through the complete build of a personal drone using the popular Little Dipper airframe.

Readers will learn to construct their own personal drone in its entirety, from initial frame build to final camera mounting. The instruction includes directions for and explanations of telemetry and the use of GPS software, while offering current real-life applications with which you can engage your drone. Plus, you'll learn about racing your personal flyer against others and the secrets of increasing your flight times.

MAKE: ELECTRONICS, SECOND EDITION
By Charles Platt : $35

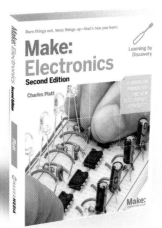

Completely re-written with most photos and schematics replaced and updated, this new iteration of Charles Platt's seminal beginner's guide to electronics continues the "learning through discovery" model for which it has been praised since the text was first published in 2009.

Single-bus breadboards are now used throughout, diagrams rather than photographs show circuit placement, internal circuit illustrations have been redrawn for clarity, and an expanded focus on Arduino as the most popular microcontroller distinguish this second edition. A mobile-friendly reading layout awaits ebook readers as well. Unchanged are the clear explanations and dry humor that have made this a best-selling book for six years and counting. Get ready to burn things out and mess things up — because that's how you learn!

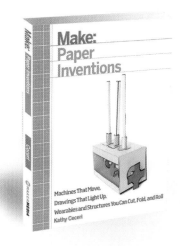

MAKE: PAPER INVENTIONS
By Kathy Ceceri : $20

Using all the many varieties of paper available, from that produced for artists to that which is fit to print the news, *Paper Inventions* explores the versatility and utility of this everyday product for a variety of applications, learning, and fun. Includes fully illustrated step-by-step instructions for 16 different projects ranging from circuits to generators, tetrahedrons to Mobius strips, baskets to sensors.

THE BEST OF MAKE: VOLUME 2
From the editors of
Make: magazine : $30

The second volume in our *Best of Make:* series, this collection gathers together the best projects published in *Make:* magazine over the last six years. This project compendium offers ideas and inspiration to expand your Maker skills.

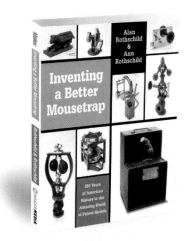

MAKE: INVENTING A BETTER MOUSETRAP
By Alan and Ann Rothschild : $25

In 1790, president George Washington signed a patent bill that recognized the right of an inventor to profit from his or her invention. But in order to receive a patent, the inventor first had to submit a small reproduction of the invention in question. Authors Alan and Ann Rothschild have made it their life's work to collect those models and showcase 4,000 of them in their own Rothschild Petersen Patent Model Museum in upstate New York. (See page 76, "Ye Olde Brushless Motor.")

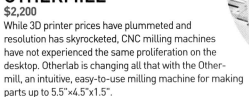

Other Machine Co.

NEW IN THE SHED

TOOLS TO BRING YOUR IDEAS TO LIFE

FILLED WITH PRODUCTS INVENTED AND CREATED BY MAKERS, the Maker Shed (makershed.com) provides everything you need to make your own projects. Be an innovator: Carve out parts with the Othermill and assemble them with PLY brackets. Test your experimental paper airplane designs with PowerUp 3.0. Then build a light show to celebrate it all with BlinkyTape.

And remember, the *Make:* team is always looking for the next breakthrough product, so if you have a kit or device you'd like us to sell, send it our way at kits@makezine.com.

Written by Michael Brady

Maker Shed
The official store of Make:

OTHERMILL
$2,200

While 3D printer prices have plummeted and resolution has skyrocketed, CNC milling machines have not experienced the same proliferation on the desktop. Otherlab is changing all that with the Othermill, an intuitive, easy-to-use milling machine for making parts up to 5.5"×4.5"x1.5".

This portable, ready-out-of-the-box tool — design-importing software included — makes 2D or 3D objects from durable materials including PVC, acrylic, copper, aluminum, wood, and even PCB. Great for circuit boards, molds, and engraving. It's perfect for anyone who's been coveting their own Han Solo-esque carbonite portrait.

USB COIN CELL CHARGER
$12

It's hard to get simpler than this: Slip your LiR2032 rechargeable coin cell battery into this charger, and plug into any computer or wall USB port. A green light on the board will let you know when it's fully charged, so you'll never have to worry about keeping your projects glowing.

Hep Svadja

PLY90 BRACKET
$30

The versatile PLY90 brackets give you a creative connector to assemble projects, no matter your experience level. Pop together a temporary desk. Prototype a brand new style of chair. Mount to the wall for a set of modern industrial shelves. Combine multiple brackets for indoor or outdoor forts. These high-strength aluminum alloy brackets hold planks of different materials, colors, sizes, and shapes without damaging them — no screws, nails, or drill needed. Easily assemble any project, and then take it apart, change it, replace panels, and customize it until it's perfect.

PLY Products

ESP8266 AND FTDI FRIEND
$7 and $10

Although it was originally designed as a Wi-Fi enabling expansion board for Arduinos, the ESP8266 can actually run Arduino code directly, making it a versatile, self-contained networking solution that happens to be crazy cheap. With powerful onboard processing and storage, as well as 6 GPIOs, it integrates with sensors and other application devices. Create an alarm to let the office know the coffee is ready, or build an accurate weather monitor to display the current conditions outside. Or simply use it to offload Wi-Fi networking functions from another processor.

You still need to build a power supply for the chip, but that's not hard — we can teach you at makezine.com/go/esp8266-microcontroller. Unlike an Arduino, it has no USB port, but the FTDI Friend is an interface adapter board that lets you connect the serial IO pins to USB — also useful for debugging.

ESP8266

Hep Svadja

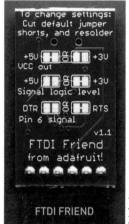

To change settings: Cut default jumper shorts, and resolder

+5U VCC out / +3U

+5U Signal logic level / +3U

DTR / RTS
Pin 6 signal

v1.1

FTDI Friend from adafruit!

FTDI FRIEND

Adafruit

BLINKYTAPE AND BLINKYTILES
$60 and $50

Blinkinlabs, makers of things blinky, has a pair of offerings that help you create your own light spectacle, for wearables, art, or ambient lighting. Each features an integrated light processor, so you don't need a separate microcontroller — just hook them up to a power source and your PC, load the Blinkinlabs software, and write code in your favorite language.

BlinkyTile is a kit of interchangeable metal shapes that you solder together to form art. BlinkyTape is, well, tape — perfect for lighting up clothing or your home theater. Each features independently controlled RGB LEDs, one per tile, or 60 on a 1m length of tape.

Hep Svadja

POWERUP 3.0
$50

Prepare to take your paper airplane to the next level with the PowerUp 3.0 — this simple device transforms it into a smartphone-controlled powered flying machine. The device adds a propeller and rudder to nearly any paper airplane, so you can customize it however you want. Sync your phone to the waterproof controller, and use the precise motion controls to fly up to 180' away. Control your newly transformed aircraft simply by tilting your phone in different directions: Tilt up to rev the throttle, and left and right to send your planes on sharp turns. It's quick, easy, remote-control flight for anyone.

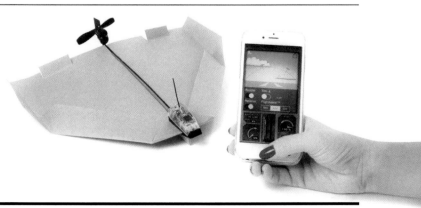

VELLEMAN VERTEX K8400

This dual-head 3D printer kit gets solid results at a low price

WRITTEN AND PHOTOGRAPHED BY MATT STULTZ

Velleman Vertex
vertex3dprinter.eu

- **Price** $799
- **Build Volume** 180×200×190mm
- **Bed Style** Cold glass with custom BuildTak sheet
- **Temperature Control?** Yes
- **Materials** Numerous
- **Print Untethered?** Built-in SD card slot and LCD makes it easy to print on the go
- **Onboard Controls?** Yes — onboard LCD with control dial
- **Host Software** Custom version of Repetier-Host
- **Slicer** CuraEngine in Repetier-Host
- **OS** Windows only for their slicer, but should be able to be used with other cross platform slicers
- **Firmware** Marlin
- **Open Software?** Yes
- **Open Hardware?** Yes, CC–BY–NC–SA 3.0

WITH MORE THAN 40 YEARS IN THE BUSINESS, VELLEMAN IS A WELL-KNOWN ELECTRONICS KIT COMPANY. In 2013, Velleman entered the world of 3D printing with the K8200, an open source 3D printer kit. Now the team has followed that with the release of the K8400 Vertex, a significant advancement from the K8200.

DUAL EXTRUDERS THAT WORK

A welcome feature of the Vertex is its optional dual Bowden-style extruders — while most DIY and kit machines have had the ability to use two extruders, Velleman has now made this an official component selection on the K8400, which is how we tested it. The Bowden setup allows both heads to be placed close to each other so the user loses less print space overall, while still having the power of dual extrusion. In our testing, we did not find uneven nozzles to cause printing problems with the Vertex. Two smoothly rotating, integrated spool holders compliment the dual extruders.

On the downside, I did find the extruders a little hard to load. I think a minor redesign of the cold end could help create a better path to make loading much easier.

NEW MECHANICALS, LARGE BED

The K8400 also introduces a re-engineered carriage system. Gone is the fixed extruder that only moved on the Z-axis on the K8200. The designers replaced it with a printhead that moves in the X and Y dimensions and a bed that moves in Z with mechanics similar to those of the community-beloved Ultimaker.

All of the linear motion components on the Vertex are built with igus polymer parts. This is a major upgrade over traditional roller bearings that need regular maintenance and have a much shorter lifespan. These polymer bearings also tend to be much quieter, a pleasant addition for in-home use.

While the Vertex doesn't sport a heated bed, the glass build plate has a custom BuildTak coating, helping with adhesion for numerous materials. The large (180mm×200mm) plate is easily removed from the printer by swiveling four metal clips out of the way. It doesn't have an auto-leveler, but its manual three-point leveling system is much easier to use than the four-point arrangements that used to be common on many machines.

CUSTOM REPETIER SOFTWARE

For slicing and printer control, Velleman has packaged their own distribution of the popular Repetier-Host. This bundle uses the CuraEngine (the back-end slicing portion of Cura) for slicing, and the custom profiles for the Vertex are bundled inside. Unfortunately, they currently only offer this package for Windows — though this is something that should be easily fixed since Repetier and Cura are both multiplatform pieces of software.

PLEASING RESULTS

The overall print quality from the Vertex was very good. The surface qualities gave quite legible text on models like the mini Makey robot. Our overhang test was also excellent. Where the printer fell down was on bridging, and any time it came to a short layer time. By default, the slicing profile has a 10-second minimum layer time. This means that on quickly printable layers like the single-walled box of the Z probe, the Vertex will move its print head away for a moment to let the layer cool. This leaves an ooze trail behind, and results in inconsistent printing when starting a new layer. This can easily be resolved by adjusting the minimum layer time settings along with the retraction settings. Hopefully Velleman will continue tweaking their profiles to optimize for this. (See our testing methodology at makezine.com/go/test-probes.)

IT'S A KIT

The Vertex, like Velleman's other products, is sold as a kit. However, the Velleman team had already assembled the unit that we received for testing. For those building their own, the company has created a very detailed set of documents. They also have an extremely active online forum with tips and tricks from other owners.

CONCLUSION

The Vertex is a great option for anyone who still wants the experience of building a kit machine. It has plenty of bonus features that make it stand out, and at its price point, it's a lot of machine for the money. ◉

PRINT SCORE: 26

● Accuracy	1 2 3 **4** 5	
● Backlash	1 2 3 **4** 5	
● Bridging	**1** 2 3 4 5	
● Overhangs	1 2 3 **4** 5	
● Retraction	1 2 **3** 4 5	
● Surface Curved	1 2 3 **4** 5	
● Surface General	1 2 3 **4** 5	
● Tolerance	1 **2** 3 4 5	
● XY Resonance	**FAIL**	PASS (0)
● Z Resonance	**FAIL**	PASS (0)

PRO TIPS

● Don't lubricate your polymer bearings. This can actually cause buildup in them that will shorten their lifespan.
● Bend your filament to straighten it and take any curl out of it when loading. This will help guide it into the Bowden tube.

WHY TO BUY

For those looking for a dual extruder 3D printer with lots of other bonus features but still want to build their own, the Vertex is hard to beat ... especially at this price point.

How'd it print?

MATT STULTZ is a software developer by day, hardware hacker by night, and an avid 3D printing enthusiast — like many others he is slowly amassing a fleet of manufacturing minions.

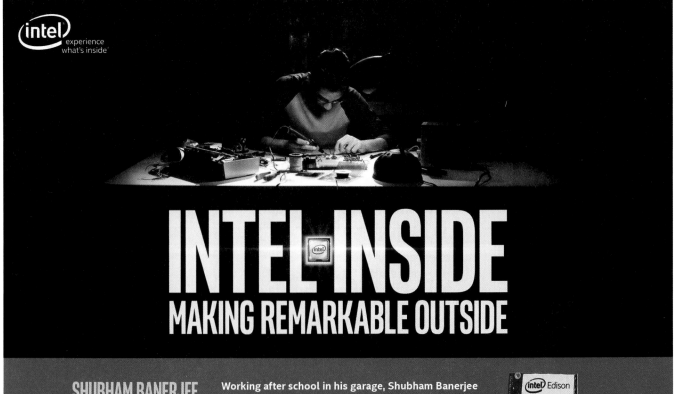

Learn to Solder with Just 1,916 Parts!

Written by James Burke

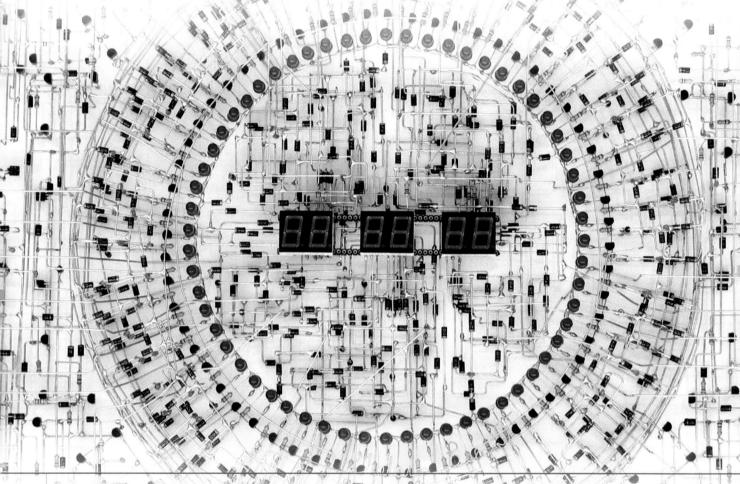

Gislain Benoit

CONGRATS ON YOUR PURCHASE OF THE LEARN TO SOLDER: MASTERS COMPONENT CLOCK KIT BY GISLAIN BENOIT. We've been very careful to include all 1,916 parts in your 14 pound shipment. On average, we advise you set aside approximately 3 years to complete this impressive build. You'll need a soldering iron, and about a dozen sets of helping hands. It's easy!

STEP 1: Begin to open and organize your kit pouches alphabetically, numerically, or by date of each component's invention.

STEP 12: Solder transistor #73T to LED #32L; repeat step for all transistors from bag "L."

STEP 14: Locate #10A resistor. Find all other #10 series resistors.

STEP 28: Yes. You found that last resistor. It wasn't at the bottom of the box was it? Under the box fold? You've got to check there first. Honestly, at this point attach it however you want, 'cause resistors don't care about polarity, so neither should you.

STEP 390: Re-tin soldering iron.

STEP 445: Look, I know component #235A was right in the middle of the table. No, I didn't move it. Why would I move it? It had to be you. You really have no business touching my property.

STEP 488: If you don't have a Ouija board, make one. Harvest wood from an ancient cypress tree located on a sacred burial ground. Be sure to coat the finished board with a wormwood-and-blood-based lacquer, or you won't get proper signal. (1/2" MDF works, too.)

STEP 578: (errata) We previously mentioned in step 400 to install the LED's anode to the transistor's collector. Please disconnect the LED from the transistor and wire it in series with the resistor. You should probably do this for all the LEDs. If you want the clock to work as a transdimensional com link, skip ahead to step #982.

STEP 675: Look, you know which one to do next. Don't ask me, I've told you like a dozen times this week.

STEP 876: If the LEDs begin to pulse in an unnerving pattern, do not attempt to interpret the blinks as Morse code. Do not decipher the blinking message under any circumstances! It will also void your warranty.

STEP 901: It's a new year. Call your parents, it's been a while.

STEP 955: (Optional) Update your Arduino IDE.

STEP 1267: Test the LEDs on your completed clock face. Plug in the clock. Smoke is not a feature, it's a problem and if detected, power should be removed immediately.

STEP 2356: Use the magnet to set the clock. The 60hz in your North American standard outlet should cycle the clock at the correct speed.

STEP 2358: Now it's time to show your friends! You can either use the dimensional communicator or the Ouija board if they've already shuffled off their mortal coil. If so, reset the clock to avoid input from alternate realities. Enjoy! ◗

Parrot
BEBOP DRONE
SKYCONTROLLER

 180°

Experience immersive flight, the future of drone piloting.

Greater distance.
Immersive controls.
More thrills!

- Lightweight and robust design built with safety in mind
- 14-megapixel "Fisheye" camera stabilized on 3-axis
- First-Person-View piloting
- Control the angle of the camera from the piloting application
- Extended range with Parrot Skycontroller add-on
- HDMI compatibility

 FreeFlight 3

 Download on the **App Store**   ANDROID APP ON **Google play**

Recommended retail price $899.99
More details on **parrot.com**

Parrot SA - RCS PARIS 394 149 496.

Always check your local laws and regulations before flying your drone. Skycontroller works with or without FPV Glasses. FPV glasses not included.
If you are using FPV glasses for an immersive flight experience, please ensure you are accompanied by a

ISBN   9781680450804